Coastal Wildlife

of British Columbia

PHOTOGRAPHY BY

Tim Fitzharris

TEXT BY

Bruce Obee

Whitecap Books

VANCOUVER / TORONTO

Whitecap Books
Vancouver / Toronto

Edited by Elaine Jones
Photographs by Tim Fitzharris
Additional photography by Graeme Ellis, pp. 3, 32, 33, 39, 40, 41, 43,
44, 45, 52; F. Stuart Westmoreland, p. 53; Fred Felleman, title page,
p. 49; Jim Darling, p. 37; Bruce Obee, pp. 36, 38, 42, 107, 114, 115,
118
Cover and interior design by Carolyn Deby
Typeset at CompuType, Vancouver, B.C., Canada
Map by Dana Dahlquist of Spring Ridge Design Ltd., Victoria, B.C.

CANADIAN CATALOGUING IN PUBLICATION DATA

Obee, Bruce, 1951-
    Coastal wildlife of British Columbia

ISBN 1-895099-86-2

    1. Zoology—British Columbia—Pacific Coast—
Pictorial works.   I. Fitzharris, Tim, 1948-   II. Title.
QL221.B7O23 1991   591.9711′1   C91-091354-4

Printed and bound in Canada by D.W. Friesen & Sons Ltd.
Altona, Manitoba

The publisher acknowledges the assistance of the Canada Council and
the Cultural Services Branch of the government of British Columbia in
making this publication possible.

Printed on 100-lb. Luna Gloss manufactured by Island Paper Mills.

Cover: Tufted Puffins.
Title page: Killer whale breaching.
Frontispiece: Snow geese populations in the northwest have
increased dramatically in recent years.
Contents page: Pintail duck in flight.

To my Parents,

Joy and Bob Obee,

for their confidence

and untiring encouragement

West Coast of North America

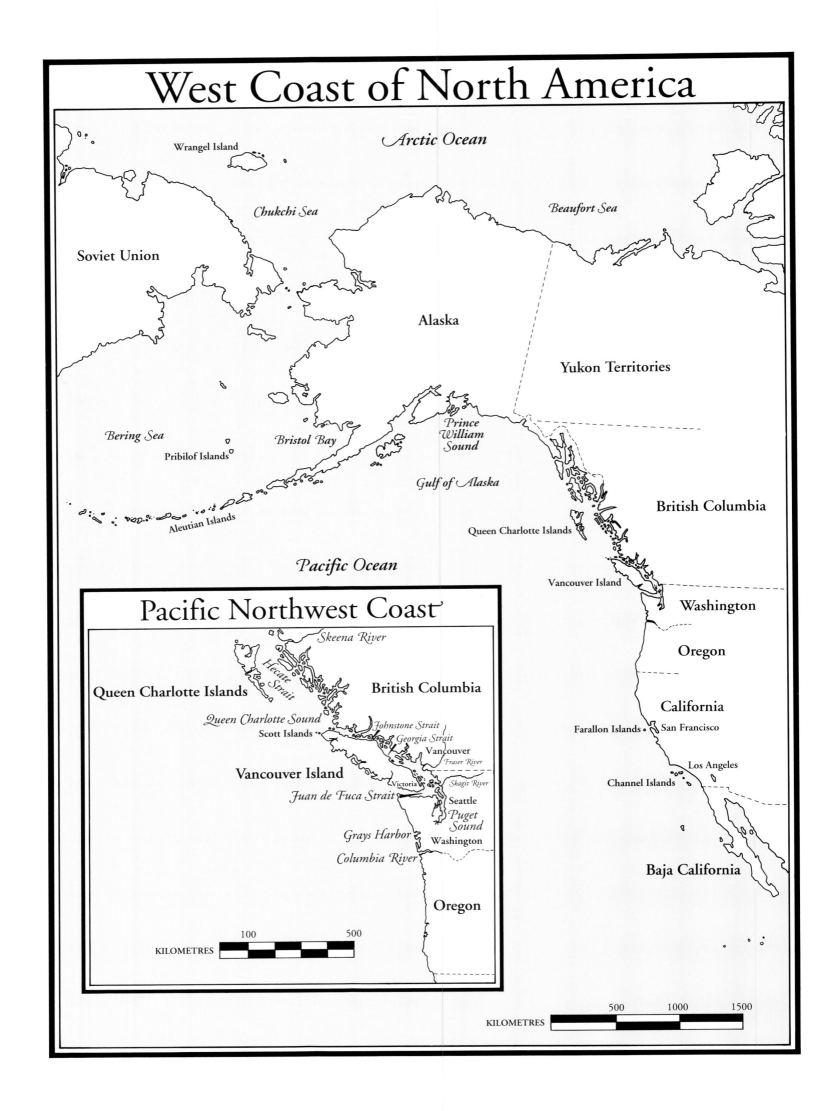

Arctic Ocean

Wrangel Island

Chukchi Sea

Beaufort Sea

Soviet Union

Alaska

Yukon Territories

Bering Sea

Bristol Bay

Pribilof Islands

Prince William Sound

Gulf of Alaska

British Columbia

Aleutian Islands

Queen Charlotte Islands

Vancouver Island

Washington

Oregon

Pacific Ocean

California

Farallon Islands · San Francisco

Los Angeles

Channel Islands

Baja California

Pacific Northwest Coast

Skeena River

Hecate Strait

Queen Charlotte Islands

British Columbia

Queen Charlotte Sound

Scott Islands

Johnstone Strait
Georgia Strait

Vancouver

Fraser River

Vancouver Island

Victoria

Skagit River

Juan de Fuca Strait

Seattle

Puget Sound

Grays Harbor

Washington

Columbia River

Oregon

100   500

KILOMETRES

500   1000   1500

KILOMETRES

C O N T E N T S

ACKNOWLEDGMENTS

Thanks to. . . *editors Bryan McGill of* Beautiful B.C., *Ian Darragh, Eric Harris, and Ross Smith of* Canadian Geographic, *for giving me a chance to write. . .manuscript readers Dr. Jim Darling, of the West Coast Whale Research Foundation, Peter Olesiuk and Graeme Ellis of the Pacific Biological Station, Ralph Archibald of the B.C. Wildlife Branch, and Ornithology Curator Wayne Campbell of the Royal B.C. Museum. . . biologists Jane Watson, Pam Stacey, Robin Baird, and Andrew Trites. . .Tony Hamilton, Andrew Harcombe, Bill Munro, and Ray Halladay of the B.C. Wildlife Branch. . .Grant Brebber of MacMillan Bloedel. . .Dave Fraser of Arenaria Research and Interpretation. . .Dr. Kees Vermeer of the Canadian Wildlife Service. . .friend and writer Rosemary Neering. . .coastal photographer Bob Herger. . .lunch mate and oceanic advisor Jim Gilbert. . .Dana Dahlquist of Spring Ridge Design. . .Whitecap designer Carolyn Deby. . .Whitecap Publisher Colleen MacMillan, for all our long-distance hours. . .editor and friend Elaine Jones. . .my wife, Janet, and daughters, Nicole and Lauren, for tolerating the grumpy writer downstairs. . .and to the memory of Dr. Mike Bigg, a true pioneer.*

Wildlife
of the
Pacific Northwest
Coast

THE ABUNDANCE AND DIVERSITY OF wildlife in the seas of the Pacific Northwest are no accident. These waters are among the most productive on earth, the lifeblood of a biological web that encompasses the most primitive and most complex life-forms. Converging currents, turbulent tide rips, eddies, upwellings, and winds combine to produce a nutrient-rich broth, fortified by minerals from the runoff of coastal

GRAEME ELLIS

streams. Phytoplankton, the simplest of all oceanic plants, is the main ingre-
dient of this broth: it blooms in the sunlit surface waters, providing the basis
for life right up to the top of the food chain.

Thousands of seals, sea lions, whales, and dolphins, millions of seabirds,
shorebirds, and waterfowl, come to feed on the prolific marine life off the coasts
of Oregon, Washington, and British Columbia. Some find year-round sustenance
here; others cross vast expanses of open ocean to stay a few months, rebuilding
their stores of fat and energy for their return to faraway breeding grounds.

Almost all the marine animals of the Pacific Northwest are seasonal. Our

LEFT: ELEPHANT SEALS LIE
TOGETHER ON THEIR
ROOKERIES.
ABOVE: SPAWNING SALMON ARE
IMPORTANT FOOD TO GRIZZLIES.

gray whales in spring belong to the Alaskans in summer, to the Mexicans in winter. The summer humpbacks of Washington and Vancouver Island are the basis of winter whale-watching industries in Hawaii and Mexico. The winter snow geese of the Fraser and Skagit river deltas are summer guests of the Soviet Union. To understand the habits and life cycles of oceangoing creatures in the Pacific Northwest, they must be viewed as inhabitants of North America's entire west coast.

Humans and animals have shared the resources of these waters for ten thousand years. The survival of coastal natives was largely dependent on seasonal migrations of aquatic birds and marine mammals, which were hunted for food, oil, clothing, tools, and utensils. Aboriginal hunting of ocean wildlife had little effect on overall populations, but with the arrival of old-world explorers, the harmonious relationship between people and animals became discordant.

Eighteenth- and nineteenth-century merchant seamen, anxious to meet world demand for furs and animal oil, brought marine mammal populations perilously close to extinction. For two centuries they slaughtered seals, sea lions, sea otters, and whales from Baja California to the Bering Sea. One species, the

Steller sea cow, was exterminated less than three decades after it was discovered by Russian traders. By the time the whaling and sealing industries had died, many marine mammal populations were mere remnants of their original numbers.

With international treaties, legislation, and decades of unmolested existence, many of these species have recovered remarkably well. For others, particularly some of the great whales, the future remains precarious.

Unmarketable marine birds fared better than animals. But during this century birds have faced a different threat from humans — habitat degradation. Nearly three-quarters of the key coastal wetlands in the Pacific Northwest are no longer of any use to water birds: they've been ditched, diked, drained, dredged, and converted to farms, neighborhoods, or industrial sites. Isolation protects some offshore seabird rookeries, but many are threatened by clearcut logging and predators introduced by people.

In coastal environments the land and sea are inseparable: they are united by the rivers that pour down from the mountains, melding with the ocean at the estuaries, the world's most productive natural habitats. In the Pacific Northwest the forests and streams above the sea are home to grizzlies and black bears, wolves, cougars, deer, racoons, and a host of other species. Many of these animals

Unrivaled Beauty

The resplendent plumage of the wood duck is one of nature's masterpieces. Iridescent green, violet, purple, and bronze; stripes, speckles, and a colorful, flowing crest; all immaculately arranged in one living package. Carvers, painters, and photographers have devoted careers to immortalizing this bird's dazzling beauty.

Wood ducks prefer swamps, flooded fields, and sleepy rivers to salt water. Most that nest in the Pacific Northwest winter in California and Mexico. Numbers of wintering wood ducks are increasing in the northwest, however, where artificial feeders attract them to parks and wildlife refuges.

Breeding wood ducks are also getting a boost from people: they almost invariably select artificial nest boxes over natural holes in deciduous trees. When the young are ready to leave the nest, they poke their heads out into the world, and take a flying leap to the ground, sometimes as far as twenty-five meters below. They literally bounce once or twice, then scamper off to the nearest water.

depend on intertidal zones, estuaries, and salmon streams for survival.

The wildlife of the Pacific Northwest coast continues to face the perils of overfishing, oil spills, pollution, entanglement in fish nets, logging, and stream degradation. Today, however, throngs of researchers are studying the effects of our industries on wildlife. On the basis of their discoveries we're developing new concepts, new management schemes, new laws to better protect our coastal and marine animals. Human encroachment and industry are here to stay, but oil-spill risks can be reduced, overfishing can be regulated, drifting plastics and pollution can be minimized, habitat damage can be mitigated. Change, however, doesn't come easily: in spite of our historic losses, the value of wildlife continues to be weighed against industrial prosperity and employment.

Coastal animals are part of the ambience, part of the quality of life in the Pacific Northwest: eagles that perch in the fir trees down the road, seals that lie on the reefs near the neighborhood beach, whales that perform for passengers aboard ferries, deer that nibble in country gardens, ducks that delight our children in city parks. It is up to everyone who admires our coastal animals to ensure their unfettered existence. We can guarantee the survival of these creatures only by protecting the environment that we leave to future generations on the Pacific Northwest coast.

Seals,
Sea Lions
and Otters

SEALS AND SEA LIONS ARE TRAVELERS,
migrants that swim thousands of kilometers
from offshore breeding rookeries to fish the seas
of the Pacific Northwest. The exception is the
ubiquitous harbor seal, a year-round resident
that is content to spend all of its days within
reasonable reach of its birthplace.

These animals are the pinnipeds, amphibious mammals that hunt at sea, but who come

ashore to bear and suckle their young. The five pinniped species of the northwest are divided into two groups — eared seals and true seals. Pinnipedia, meaning "feather" or "fin-footed," is a reference to their flippers. While all have streamlined bodies, like blimpish tubes with tapered ends, the flippers vary. The eared seals — fur seals, and California and Steller's sea lions — have tiny, leatherlike external ears. Their fins are enormous, enabling them to move with surprising agility on land. Their hind flippers rotate under their bodies to provide momentum. Their front flippers are long enough to hold their bulky bodies upright. Underwater they are used like wings, for both propulsion and steering.

The true seals — harbor and elephant seals — hear through small holes in the sides of their heads. Their short foreflippers are virtually useless on land: they lie prostrate, awkwardly worming along on their bellies to reach the sea. But in the water they move gracefully, propelling themselves with hind flippers, steering from the front.

Fur seals and sea lions are believed to have evolved from the same ancestral line as bears. Harbor and elephant seals were more closely related to weasels. All of the pinnipeds have fur and whiskers, but they rely on thick layers of blubber for warmth, buoyancy, and energy storage.

Like seabirds and whales, seals and sea lions are equipped for long, deep dives. Their muscles contain high levels of myoglobin, a respiratory pigment that allows them to store oxygen. By exhaling before submerging, they slow the heartbeat; peripheral blood vessels contract, saving circulation for the heart and brain.

Another amphibious mammal, the sea otter, is not as common in the Pacific Northwest. A member of the family Mustelidae, this furry-faced creature is a

PREVIOUS PAGES: FUR SEALS ON BREEDING ROOKERY IN ALASKA'S PRIBILOF ISLANDS. INSET: ONLY MALE ELEPHANT SEALS HAVE THE DANGLING PROBOSCIS.

BELOW: ABOUT FOUR THOUSAND CALIFORNIA SEA LIONS WINTER IN THE PACIFIC NORTHWEST.

BRUCE OBEE

relative of the mink, badger, or skunk. Biologically, the sea otter is much like its closest cousin, the river otter: it is thought that both species evolved from Indian and eastern Asian stock and moved north and west to North America.

All of these marine mammals make delectable dinners for killer whales and, south of Washington, great white sharks. Since the mid-eighteenth century their most worrisome predator has been man: many of these species were reduced to near-extinction before international agreements and legislation finally protected them. Though no longer legal quarry, they nonetheless remain largely at the mercy of humans. As they forage the seas today, they face the hazards of chemical effluent, monofilament gill nets and drift nets, oil exploration and transportation, and destruction of habitat from industrial and residential development. Natural climatic and oceanographic changes may also have an adverse effect on the future of these animals. While many marine mammal species appear to have stable or growing populations in the Pacific Northwest, there are recent, and drastic, declines in the north.

Sea Lions — Little Fear of Man

When the sea lions return to Vancouver Island each fall, tour boats from the city of Nanaimo run down Northumberland Channel to the lumber and pulp mills at Harmac. As many as two thousand sea lions haul out on log booms and lie burping and barking, noses in the air as if posing for the cameras. They seem oblivious to boom boats puttering around them, rearranging logs while trucks and workers bustle above the shore.

Most of these blubbery beasts are California sea lions that have migrated more than two thousand kilometers from southern breeding waters. They represent about half the California sea lions that winter in British Columbia, along with larger Steller's, or northern, sea lions. They feed on big schools of herring that spawn in Northumberland Channel. Bald eagles, seals, and seabirds join the sea lions for a feeding frenzy that lasts from December to March. When the spawn is over, in April or May, the sea lions begin the long swim to their summer homes.

The population of California sea lions wintering in B.C. and Washington has exploded since the 1970s, and their use of the Harmac log booms is an example of their affinity for man-made haulouts. Across Georgia Strait, as many as seven hundred take over the Sand Heads jetty in spring when the eulachon are running up the Fraser River. In Puget Sound, they climb aboard log booms, barges, construction rafts, and oil rigs at Port Gardner, and Elliot and Shilshole bays. Large groups raft together and drift in Elliot Bay, beneath the imposing skyscrapers of Seattle. Since 1980 they've wreaked havoc on steelhead trout returning to the Lake Washington system through Seattle's Chittenden Locks. Farther south they follow the eulachon up the Columbia River, and chase lamprey at the mouth of Oregon's Rogue River and the Klamath River in California. Since the late 1980s, California sea lions have been a tourist attraction at San Francisco's Pier 39, where they have taken over a marina, forcing yachtsmen to find new mooring spots.

Sixty years ago, after more than a century of commercial exploitation, California sea lions had fallen to an all-time low: by some estimates there were as few as a thousand left. Today breeding rookeries between the Gulf of California and the Farallon Islands, off San Francisco, may support about 160,000.

There is probably a comparable number of Steller's sea lions, breeding in rookeries along the North American coast from the Channel Islands, off Los Angeles, to the Bering Sea. There are concerns, however, about declines in Steller's populations, particularly in the Pribilof and Kodiak islands of Alaska.

In British Columbia, Steller's sea lions mate at rookeries on the Scott Islands, off the northwest tip of Vancouver Island; at Cape St. James, off the south end of the Queen Charlottes; and at North Danger Rocks in Hecate Strait. The breeding population appears to have stabilized at about six thousand, which produces about twelve hundred pups a year. At a rookery on Alaska's Forrester Island, less than twenty kilometers north of the B.C. border, about twenty-three

hundred pups are born each year.

The mating rituals of both California and Steller's sea lions are a grueling test of strength and endurance for the bulls. To an observer who can tolerate the old-fish stench, the scene on a rookery is chaotic, with thousands of sea lions barking, biting and battling. Mature Steller's bulls, the largest of the eared seals, measure more than ten meters long and weigh a whopping tonne. They are more than three times as large as their female companions. And a full-sized, 360-kilogram California sea lion bull is not an especially gentle partner for a 110-kilo female of the same species.

Steller's Sea Lions

The Steller's, or northern, sea lion is the largest of the eared seals in the northwest, with bulls weighing up to a tonne.

When a heavy surf is running during stormy weather, Steller's sea lions prefer to stay in the water. They form compact groups and bob about in the sea a few meters from shore. On clear days when winds are light, they haul out on beaches and reefs to bask in the sun. During the day they seem to spend more time on land in the afternoon.

Although leery of people, many Steller's sea lions will hold their ground when approached, taking refuge in the ocean only if they feel threatened. Canoeists and kayakers often paddle among swimming sea lions, and scuba divers sometimes join them underwater.

A breeding bull stakes out a territory and defends it vehemently: intruding bulls are met by hissing, snarling, neck-bashing and biting. Once the dominance is established, the master may stay put, without food, for as long as two months. Females aren't particularly faithful and seem to remain within territorial boundaries more to avoid warfaring males than to be part of a specific mating team. A single bull might mate with thirty females, and each cow will bear one pup.

The annual return of the sea lions brought early-day sealers to the rookeries. The choicest animals were herded away from the sea and selectively slaughtered, usually clubbed or shot to death. The blubber of three or four sea lions could yield one barrel of oil. Skins were sold for various uses, testicles became Chinese medicine, and whiskers adorned many nineteenth-century hats. Alaskan natives

used the oil in lamps, ate the flesh, stretched skins over boat hulls, and twisted sinews into thread. They sewed the throat lining into boots, used fins for shoe soles, made intestines into waterproof clothes, and inflated stomachs for floats and oil bottles.

In late summer or early fall the rookeries are abandoned. Adult females with young don't travel far, but juveniles and males disperse and often swim thousands of kilometers away to fatten up in preparation for the rigors of the next season. Animals seen long distances from breeding rookeries are usually juveniles and bulls.

Where California and Steller's territories overlap, the differences between species may not be obvious to casual observers. The California sea lion, besides a smaller body, has darker fur, often black, while a Steller's is auburn, sometimes

ABOVE: CALIFORNIA SEA LIONS
ON LOG BOOMS AT HARMAC
LUMBER AND PULP MILLS.

lighter. The California sea lion is the well-known "circus seal" that balances beach balls and performs at oceanaria around the world. It's the one with the sharp bark that can be heard across vast stretches of calm, open water. Steller's sea lions appear more ornery, and are likely to growl at unwary sightseers. One of the most distinguishing features appears on mature Californias — a prominent bump or forehead, and a white crest, above the long snout.

While wintering away from the rookeries, sea lions face other adversaries — fishermen. A sea lion may dive to 140 meters, stay down fifteen minutes at a time, and gobble twenty kilograms of herring, salmon, octopus, squid, halibut, hake, and dozens of other fish every day. Fishermen have argued for the past century that the only good sea lion is a dead one. Governments have responded with "control programs."

In Canada a bounty on their snouts had little effect in reducing populations, so in the early 1920s fisheries officers shot them on the rookeries, at their feeding areas, and at haulouts. In the late 1960s, environmentalists rallied against the killing and convinced the federal government, in 1970, to protect marine mammals under the Fisheries Act. Similar protection was given in the United States two years later.

The controversy, however, is far from over. Fishermen, who lose gear worth hundreds of thousands of dollars to sea lions every year, continue to call for controls. A few take the law into their own hands and shoot sea lions from their boats. Meanwhile, some researchers argue that sea lions prey on hake, lamprey, and other species that are harmful to commercial fish. Sea lions, therefore, could be allies, rather than enemies, of the fishermen.

Fur Seals — Offshore Winter Visitors

Northern fur seals are abundant off the Pacific Northwest coast in winter, yet they are seen infrequently from shore. Unlike harbor seals, sea otters or gray whales, which feed in near-shore waters, fur seals often congregate over the edge of the continental shelf. The outer edge of the shelf lies along the two-hundred- or three-hundred-meter contour, anywhere from a few hundred meters to 120

Fur Seal Convention

Between the mid-1700s and early 1900s, perhaps as many as 5 million fur seals were killed in the North Pacific. By the early twentieth century the populations of fur seals breeding in the Pribilof Islands had been reduced to fewer than 300,000 animals.

To stem the decline, Russia, Japan, Great Britain (for Canada), and the United States agreed to ban killing on the open ocean. The North Pacific Fur Seal Convention of 1911 compelled sealers to hunt selectively on land, and established sharing arrangements for the skins. The treaty expired in 1940 and, in 1957, the same four countries negotiated the Interim Convention on Conservation of North Pacific Fur Seals. The agreement has been amended and renewed over the years. Today only aboriginal subsistence sealing is allowed in the Pribilof Islands.

kilometers offshore. Out here, fur seals dine on herring, salmon, walleye pollock, squid, eulachon, or rockfish. They are particularly fond of schooling fish, and though fur seals can dive to two hundred meters, they generally feed at night when their prey moves closer to the surface.

Fur seals wintering in the Pacific Northwest are migrants that disperse from northern breeding rookeries in autumn. Most bulls winter south of the Aleutian Islands and in the Gulf of Alaska. Females and young travel south, some as far as California's Channel Islands. They are most plentiful off Washington and British Columbia from December to April or May. Some meander into Juan de Fuca Strait or Puget Sound, but few venture within fifteen or twenty kilometers of shore. They rarely leave the water, except to mate.

Native Indians paddled far offshore to hunt these seals: anthropologists say that for two thousand years before the turn of the century, 80 percent of the mammal remains in middens at Cape Alava, Washington, were those of fur seals. After Europeans arrived, wintering seals were the mainstay of a thriving sealing industry. Victoria, B.C., only 150 kilometers from Cape Alava, was an important base for pelagic sealers in the late 1800s and early 1900s.

A big bull fur seal might weigh 250 kilograms — five times as much as a large cow. In spite of their bulk, bulls are surprisingly quick when defending their turf on the rookeries. It was on a northern rookery that these burly beasts were first described by naturalist Georg Wilhelm Steller, who called them "sea bears." Steller accompanied Vitus Bering, of the Russian Navy, on the ship *St. Peter*, which was wrecked in the winter of 1741 on Bering Island, off Kamchatka Peninsula. Steller and others who survived the pitiless winter built a lifeboat from the wreckage as the weather began to warm. While they were building the boat, the first of the mating fur seals returned to the island.

Steller, however, had discovered a minor colony and it was his news of the supposedly more plentiful "sea beaver" — or sea otter — which launched Russia into the profitable North Pacific fur trade. But in 1786 Soviet fur hunter Gerasim Pribilof stumbled upon the islands which now bear his name, and discovered the world's largest population of fur seals.

When the animals were most plentiful on the rookeries, from July to September, early sealers staged "knock-downs" in which hundreds of seals were killed. They were herded inland, as far as ten kilometers from the breeding grounds. Seventy or a hundred bearing the finest skins — three- and four-year-olds — were separated from the flock and dispatched with a sharp blow to the snout. Oil was boiled from the blubber; skins were salted in bins and packed for shipping. Valued at about five dollars each in the late 1800s, skins were also bought from natives for as little as ten cents apiece.

Although the methods of these early sealers may seem gruesome, they were at least selective. On the high seas, hundreds of thousands of fur seals were killed indiscriminately as they migrated to and from the rookeries. During a twenty-year period around the turn of the last century, about 600,000 of these migrating seals, mainly cows, were killed.

Before Russia sold Alaska to the United States in 1867, well over 2 million fur-seal skins were harvested from the high seas and rookeries. By 1911, when fur-seal hunting was limited by international agreement, another 3 million were killed.

Like other marine mammals, fur seals recovered from the exploitation of previous centuries. By the mid-1980s the world population was more than a million, with some 900,000 using rookeries in Alaska's Pribilof Islands, 330,000

19

PREVIOUS PAGES: HARBOR SEALS, CAUTIOUS BUT CURIOUS, OFTEN SURFACE NEAR SMALL BOATS.

on Soviet islands, and about 3,000 on San Miguel Island in California.

As we enter the 1990s, however, fur seal populations appear to be diminishing. A common suggestion is that their insatiable curiosity gets young seals entangled in abandoned fish nets and they drown or choke to death. A more recent theory is that fisheries mismanagement has caused drastic declines in prey species and northern marine mammals are starving to death.

Harbor Seals — Year-Round Residents

With so many harbor seals near the cities of the Pacific Northwest, it's almost impossible to go to a beach and not see one. They skulk along with noses just above the surface, shiny gray heads glistening like glass fishing floats. Cautious but curious, they float silently, watching the movements of beachcombers through bulging black eyes.

On ebb tides, harbor seals gather on low-lying reefs and shelves, or haul out on docks, sandbars, and log booms. Unlike a sea lion or fur seal, a harbor seal's front flippers are too short to prop up its lumpish, cigar-shaped form. They lie at the edge of the water, ready to bounce awkwardly along on their bellies and dive into the sea at signs of approaching danger.

When haulouts are not within easy reach, harbor seals rest on the bottom of the sea for five or six minutes, come up to breathe for a moment, then return to the seafloor to snooze. They can dive deeper than 180 meters and stay submerged for twenty minutes.

Harbor seals are quiet creatures, emitting low-pitched grunts, perhaps a form of communication. They are snoopy, and canoeists or kayakers are often startled by the watery snorts of seals surfacing a paddle stroke or two from the boat. They do not bark: the so-called barking seals of circus fame are actually California sea lions.

Harbor seals vary in color, from black to white with shades of bluish gray. Most have spots, or mottled patches on their backs. When full grown, both sexes are of similar size, weighing sixty or eighty kilograms and measuring about 1.5 meters from nose to hind flippers. A male may live twenty years, ten years less than a female.

These are nonmigratory animals: almost all of their activities center around their chosen haulouts. Breeding occurs from April to September throughout their territories, rather than on specific rookeries. There is a lot of excitement at mating time, with couples splashing on the surface, biting one another, jumping clear out of the water. They copulate in the water and the mother is left to give birth and care for its single pup alone. Like sea otters, harbor seal cows form strong bonds with their offspring and will attend a dead pup for several days. Feeding on milk with a fat content of more than 40 percent, a ten-kilogram newborn will double its birth weight in five or six weeks. In urban areas pups temporarily left alone are occasionally thought to be abandoned, and poorly informed good samaritans carry them off to humane societies.

Groups formed by harbor seals don't appear to be bonded by any family ties: they are simply animals occupying the same area. Gatherings are small, usually fewer than fifty, but herds of six and seven hundred congregate at river mouths to whelp and to catch fish moving in to spawn. Harbor seals travel long distances upstream on incoming tides: they have been seen a hundred kilometers up the Columbia River when the eulachon are running, and are known to chase fish into Pitt Lake, fifty kilometers from the mouth of the Fraser.

On an average day, a full-grown harbor seal will devour two or three kilograms of fish. They are opportunists, feeding on hake, herring, salmon, lingcod, squid, sculpins, and whatever else is available. In response to complaints that seals were cutting into fishermen's incomes, biologists in the late 1980s studied feeding habits of seals in Georgia Strait. Hake and herring, both commercial fish, comprised three-quarters of their diet. Salmon, however, made up only 4 percent.

Seals in the Canadian Gulf Islands and American San Juans are crafty hunters: they've learned how to acquire easy meals by hanging around favored sport-fishing spots. When a salmon strikes an angler's lure, seals are alerted by the sound of fishing line zinging out from the reel. They slide up behind the boat and snatch the prize. Some sport fishermen claim to lose sixty or seventy fish a year to seals.

Anglers have recently joined commercial fishermen in calling for government-sponsored seal kills to reduce competition. In Canada, federal government biologists say about twelve thousand harbor seals could be culled each year from British Columbia's current population of about a hundred thousand.

On the coast of North America, Pacific harbor seals are found from Mexico to the Bering Sea. Healthy harbor seal populations in the Pacific Northwest are the result of legislation passed in the early 1970s which gave them full protection. There were probably ten thousand harbor seals left in B.C. before 1970. Bounties claimed an average of three thousand seals a year between 1928 and 1964; an equal number were probably killed, but sank before they could be retrieved. A commercial hunt in the late 1960s also eliminated about ten thousand harbor seals. Since they became protected in B.C., their numbers have grown

BELOW: HARBOR SEALS MOVE AWKWARDLY ON LAND.

by about 12 percent a year. In Washington, where bounties claimed fifty-two thousand seals from 1940 to 1960, harbor seals are also thriving.

Elephant Seals — Mammoth Mammals

The elephant seal is the largest pinniped in the northern hemisphere. Twice the size of a Steller's sea lion bull, a full-grown male weighs two tonnes and measures five meters from its dangling proboscis to its hind flippers. A cow is three meters long and weighs a tonne, but does not have the peculiar trunk hanging over its muzzle.

These mammoth mammals are part-time residents of the Pacific Northwest, arriving in April, staying until June or early July. Elephant seals off Oregon, Washington, and British Columbia are usually adult bulls and juveniles that migrate north from their winter breeding rookeries in California and Mexico.

Populations are growing and sightings in the northwest are increasing: winter sightings now are becoming common. Some travel into Puget Sound but they are seen more often on the west coast of Vancouver Island, in Hecate Strait, and in southeast Alaska. Like sea lions, elephant seals were once important to Indians in these areas.

When away from the rookeries, elephant seals are solitary animals, rarely coming ashore except to molt. In the water the elephant seal may look like a grayish-brown deadhead floating menacingly in the course of passing boats. A skipper steering around one may be surprised to see the "deadhead" suddenly disappear.

At sea these seals are deep-water hunters. They travel hundreds of kilometers offshore: fishermen see them sixty or eighty kilometers out, hunting over the continental slope. Dives to three hundred meters are common, and elephant seals are known to reach depths of six hundred meters or more. They spend almost all their time underwater. Their routine, week after week, is to remain submerged for half an hour, come up to breathe for two or three minutes, then dive again. Sometimes they stay beneath the surface for two hours. Sealers who found pebbles the size of golf balls in their stomachs thought elephant seals took on ballast before going down. These seals get along with fishermen better than other species: they take ratfish, dogfish, skate, squid, cusk eels, and other fish of little commercial value.

Nineteenth-century sailors carved pipe bowls from elephant seals' teeth and used pelican bones for stems. The seals were valued for their oil, which was superior to whale oil as a lubricant. One gigantic bull taken at California's Santa Barbara Island in 1852 was 5.5 meters long — 795 liters of oil were boiled from its blubber. Hundreds of thousands of northern elephant seals were slaughtered for their oil, and by the turn of the century only a few hundred remained. Today there may be 100,000 on the Pacific coast.

Otters — Repopulating Old Domains

Sea otters forage undisturbed along the reef-ridden shores of the North Pacific, as they did long before the arrival of old-world explorers. Thought to be nearly extinct less than a century ago, these endearing clowns of the sea are reappearing through much of their traditional domain. They are most at home in kelp beds, lying bellies to the breeze, like fuzzy-faced old men with handlebar whiskers, strawberry-sized black noses and curious, frowning eyes. They gather in groups and wrap themselves in kelp to prevent drifting while asleep.

Although they can dive to ninety meters, they generally feed in shallow water on urchins, abalone, crabs, and other shellfish. Underwater, otters store food in pouches of loose skin under their left forearms. They are the only marine mammals to use tools: shells too hard to crush in their powerful jaws are placed on their chests and pounded with rocks held in their forepaws.

The otter's eating habits cause major ecological changes. At Checleset Bay on northwest Vancouver Island, otters feed on red sea urchins, shellfish which rasp along the seafloor, inhibiting the growth of kelp. Until the reintroduction of sea otters in the late 1960s and early '70s, there were few kelp beds at Checleset Bay: today vast forests of kelp are new nurseries for young fish, snails, and crabs.

The otter is the smallest marine mammal — with females weighing about thirty-two kilograms, males about forty-five — yet it is a large consumer. On

Mistaken Identity

Young male elephant seals, yet to develop their bulbous noses, and great size, can be mistaken for sick harbor seals when molting. Like adults, they haul out on beaches and lie there for days as patches of outer skin and hair gradually peel from their bodies, causing minor infections and bleeding. Their continuously watering eyes give the impression that they're suffering. During this process they seem unafraid of people who approach, and even touch them. Only an expert can differentiate between a harbor seal and young elephant seal, and molting juvenile elephant seals are occasionally "put out of their misery" by well-meaning, but ill-informed animal lovers.

BRUCE OBEE

an average day an adult sea otter may eat six or eight kilos of meat — about one-fifth of its body weight. The sea otter's voracious appetite has put it at odds with fishermen, who generally oppose transplants of otters to areas inhabited by commercially valuable shellfish.

Humans have always been the otter's worst enemy. Historically the species wasn't a competitor, but a commodity, an abundant source of what explorer Meriwether Lewis called "the most delicious fur in the world." Old-world merchants first learned of this unexploited wealth in the eighteenth century. Vitus Bering, of the Russian Navy, was shipwrecked off Kamchatka Peninsula in 1741 when he was returning home with news that he'd discovered Alaska. Bering died but the survivors built a lifeboat and safely sailed to Russia. One survivor was Georg Wilhelm Steller, a naturalist who brought back otter skins along with the world's first description of the "sea beaver."

Russian fur traders had a burgeoning industry in the North Pacific by 1778, when Captain James Cook, of the British Navy, stopped at Alaska's Prince William Sound and traded iron nails for sea-otter furs. "The fur of these animals is certainly softer and finer than that of any others we know of," Cook's log notes, "and therefore, the discovery of this part of the continent of North America, where so valuable an article of commerce may be met with, cannot be a matter of indifference."

Soon it was the sea otter, not the mountains of timber or gold-filled streams, which lured early explorers to North American shores. They were eager to satisfy

ABOVE: RIVER OTTERS ARE
ABUNDANT ALONG NORTHWEST
SHORES.

the wealthy, fashion-conscious mandarins of China and the aristocracy of Europe. Capitalizing on native hunting skills, they traded with the Aleuts of Alaska and Indians of the northwest coast — otter skins for iron tools and nails, copper kettles, pewter jugs and tankards, guns and gunpowder.

By the late 1700s and early 1800s, dozens of ships were hunting the lucrative sea otter throughout its range, from Baja California to the Bering Sea. Some reports say Canton alone received nearly 50,000 pelts from 1799 to 1802, and at least 200,000 more went on the world market within the next decade.

But the bloody prosperity couldn't last: in 1900 fewer than 130 skins were gathered between California and Alaska. Eleven years later, when Russia, Japan, Great Britain (for Canada), and the United States agreed to stop otter hunting, the total world population was estimated at between 1,000 and 2,000 animals. Some historians say nearly 370,000 otters were taken from the North Pacific

before 1911; others believe it was nearly a million. The only certainty is that prospects for the species were bleak.

For two decades, sea-otter sightings were rare, but in the mid-1930s they began to reappear. A hundred were spotted near Monterey Bay, California, a population which now has reached about 2,000. In the 1940s thousands were known to once again inhabit Amchitka Island, at the western end of the Aleutians. In Alaska today, there are well over 150,000, occupying virtually all of the sea otter's historic domain. Otters transplanted from Alaska have seeded new populations along the coasts of Vancouver Island and Washington.

Rebuilding these populations is a lengthy task. An otter, which may live for twenty years, reaches sexual maturity at about three years. After mating in water, the female abandons its partner and later gives birth in the sea. More than one pup is uncommon and females breed every year or two, depending on food supplies.

The bond between mother and pup is legendary, so powerful, in fact, that mothers have been known to cling to dead offspring for several days. A young sea otter spends many of its early days comfortably cuddled on its mother's chest. When the mother dives for food the pup drifts quietly on its back. Several times a day the mother methodically preens the fur of its young, then its own, ensuring continued protection from frigid seas.

In 1970 sea otters became protected in Canadian waters under the Fisheries Act. Americans followed two years later with the Marine Mammals Protection Act. Though protected by law, the life of a sea otter today is not without its perils. Killer whales and sharks sometimes dine on otters, or pups are snatched away by eagles, but the sea otter's most formidable foe is still the human race with all its industries. An estimated fifty-five hundred sea otters were killed in

Prince William Sound after the tanker *Exxon Valdez* ran aground in March, 1989. It was a lamentable loss, although not threatening to Alaska's abundant supply. However, if a similar spill occurred near the newly established colonies in B.C. and Washington, or farther south in California, the sea otter could once again be extinguished in parts of its southern range.

The sea otter's closest relative, the river otter, could be inappropriately named. While it does occur in inland lakes and streams, it is one of the most abundant seashore mammals on the Pacific Northwest coast. Although its fur is inferior to that of the sea otter, the river otter was hunted by early traders and remains the legal prey of modern-day fur trappers. An adult male river otter weighs about fourteen kilos, less than half as much as a sea otter.

Known for their gregarious nature and playfulness, groups of four or five river otters enjoy frolicking in the snow, or sliding down mud banks and slippery rocks. Underwater they swim fluidly, undulating as though their bodies are boneless. They search shallows and tide pools for fish, crabs and shellfish, leaving evidence of their hunting success in droppings laced with fish remains and crushed shells.

Although entertaining to wildlife watchers, river otters can be the bane of cottagers and yachtsmen. Damp crawl spaces or cozy cockpits make comfortable dens to give birth or store dead fish.

Like sea otters, river otters don't migrate long distances, but they do appear to have well-developed homing instincts. A pair taken from Victoria Harbor to Vancouver's Stanley Park Zoo escaped: within three weeks they traveled 115 kilometers, including nearly 25 kilometers across Georgia Strait, to their Victoria home.

The river otter is often confused with another coastal creature, the mink, a member of the same biological family. The weasellike mink, however, is much smaller than the river otter, weighing only about one kilogram.

Successful Second Chance

Sea otters, hunted to near extinction by fur traders, had disappeared from most of the North American coast by the early twentieth century. They were protected by international agreement in 1911, and by 1940 populations in Alaska were showing noticeable improvement. In the late 1960s and early '70s Alaskan sea otters were transplanted to vacant habitats. More than four hundred from Prince William Sound were released in southeast Alaska. Eighty-nine transplanted from Amchitka Island and Prince William Sound to Checleset Bay, on northwest Vancouver Island, have proliferated to about six hundred: they are gradually spreading along the outer coast of the island. Thirty otters released at Point Grenville, Washington, didn't fare as well, but another thirty transplanted farther up the coast, at La Push, have been more successful: now there may be two hundred sea otters in Washington. Ninety-three otters translocated to Oregon disappeared. While sea otters don't migrate, they move to new areas as populations grow and deplete food sources.

GRAEME ELLIS

GRAEME ELLIS

Whales, Dolphins and Porpoises

FOR CENTURIES THE GREAT WHALES have captured our imaginations simply by breaking the surface to breathe. It's their sheer immensity, as they rise from the ocean depths and burst into view, plumes of mist exploding from their blowholes. They puff like locomotives, then roll headlong into the sea, swallowed by the ocean from which they so suddenly appeared.

It's hard for terrestrial beings like ourselves to envision these aquatic behemoths roaming their undersea domains. Only when they come to share our atmosphere do we sense some kind of mammalian bond between human and whale. In recent years this kinship has lured scientists to sea, eager to understand the myths and mysteries that have perplexed us for so long. We've learned of their intelligence and sensitivity, of their remarkable maternal instincts, of their extraordinary migrations, of their songs and communications. Yet it seems the more we learn, the more we see how little we know about whales. Population sizes of many species are unknown and estimates of others are merely educated guesses. The discovery of new populations raises more questions. Where do they come from? How extensive are their ranges? How many other whales are yet to be found in the North Pacific?

Another unanswered question is how well these magnificent creatures have recovered from abuses of the past. Thousands upon thousands fell victim to the harpoons. Sperm whales, blue whales, minkes, fins, seis, humpbacks, grays, right whales, bowheads, and more felt the pain of the grenades exploding within their bodies. Whalers from Norway, England, America, Canada, Russia, Japan, and many other countries came to exploit the whales. The old Malay proverb — "However big the whale may be, the tiny harpoon can rob him of life" — proved true in the North Pacific. Although no one knows precisely how many whales once roamed the sea, it is widely believed that with a few exceptions — such as the Pacific gray whale — today's whale populations are mere remnants of traditional numbers. It seems a surety that most great whale species will never again inhabit the earth in their historic abundance.

Today, however, with international agreements and laws to protect whales, whaling is more endangered than some of the species it claimed. Few whales are taken commercially, and the hunting of small numbers is permitted for aboriginal use. Our interest in whales now is one of conservation and research.

Whales, or cetaceans, are divided into two groups, baleen and toothed whales. The nine baleen whale species in the North Pacific range from the ten-meter-long minke, to the blue whale, at twenty-six meters long the largest animal on earth. The blowholes of these whales are divided in two, producing a V-shaped spout when they exhale. Depending on the species, they have as many as eight hundred plates of baleen, or whalebone, hanging from their upper jaws. When feeding, they suck in great gulps of water laden with fish, plankton, krill, and other tiny invertebrates. Using the force of their enormous tongues, they expel the water through the baleen plates, which trap the food. Although they can dive to more than 350 meters and stay submerged almost an hour, many baleen species prefer to feed within fifty meters of the surface, coming up for air at least every five or ten minutes.

Toothed whales, about thirty species in the North Pacific, include dolphins and porpoises. They have single blowholes and anywhere from 2 to 250 teeth. It is not unknown for toothed whales to dive as deep as 450 meters and stay down more than two hours. The largest of this group, the twenty-one-meter, fifty-tonne sperm whale, is reportedly capable of diving as deep as three kilometers.

A number of physiological factors allow whales to attain such great depths for so long. While humans may use their lungs to 20-percent capacity, whales can fill them to 80 or 90 percent. Humans absorb about 20 percent of the oxygen they inhale, compared to 50 percent in whales. Myoglobin, where oxygen is stored, is up to nine times higher in the muscles of whales than in land mammals. Like seals and sea lions, the heartbeats of whales slow down on deep dives, saving circulation for the most vital organs, the heart and brain.

Many whales, particularly the large baleens, are plagued by parasites. Each whale has its own particular brand of barnacles which attach themselves to the skin. Lampreys, suckerfish, and others affix themselves to the side of a whale and travel wherever the whale goes. These parasites rarely cause fatal or serious infections.

The toothed whales have built-in sonar systems. They transmit repeated clicks and other noises, which bounce off objects, helping locate food, enemies, or other whales. Their echolocation abilities, however, are insufficient to detect large-mesh nets, and each year several whales are inadvertently drowned by fishermen. Fine monofilament nets are particularly lethal for dolphins: tens of thousands are killed by Asian driftnet fishermen who are fishing for North Pacific squid. But their methods are indiscriminate, tantamount to strip-mining the sea. In the past, tuna fishermen have also taken large incidental catches of dolphins, which habitually swim above schools of tuna. During recent years, however, new tuna-fishing methods have been developed to minimize, even eliminate, the accidental killing of dolphins.

Dolphins and porpoises that survive the perils of the sea may live thirty, forty, even fifty years. Wild whales may live fifty or sixty years — a real old-timer could reach eighty.

Pacific Gray Whales — Back from the Brink

Between late February and early May, about twenty-one thousand Pacific gray whales migrate through the Pacific Northwest to summer feeding grounds in the north. Most travel up the outer coast, but a few usually stray into Juan de Fuca and Georgia straits, or mosey down into Puget Sound. These inside-water whales have been seen amid the traffic in Vancouver Harbor and in the lower reaches of the Fraser River. They hang around all summer, then join the southbound migration some time between November and early January.

The incredible sixteen thousand-kilometer return migration of the Pacific, or California, gray whale is one of the longest of any mammal on earth. Traveling alone, in pairs and trios, or in groups up to sixteen, the first migrants move

ABOVE: YOUNG GRAY WHALE
NEAR VICTORIA HAS STRAYED
INTO INSIDE WATERS.

north from their breeding lagoons in Baja California in late January. Pacific grays, unlike killer whales, do not form pods, or close-knit families that stay together for life.

Often within a hundred meters of shore, the whales are prime prey for naturalists. At Long Beach, in Vancouver Island's Pacific Rim National Park, the migration peaks around Easter. About forty or fifty remain off Vancouver Island through summer and fall, then join the southbound migration. These resident whales, along with the migrants, are the mainstay of a thriving whale-watching industry. One feeding whale may attract hundreds of boaters, kayakers, and airplanes for weeks at a time.

Whales migrating past Long Beach don't stop to feed, but travel just beyond the surf line. Naturalists, dressed in winter woolies and rain gear, perch on headlands and scan the sea for the heart-shaped plumes of surfacing whales. Sometimes a whale will breach or raise its flukes out of the water. More often observers see just the row of bumps, or "knuckles," on the lower back as the whale blows, then quietly rolls headfirst into the depths.

Beyond the Pacific Northwest, migrating grays hug the shoreline into the Gulf of Alaska and Bering and Chukchi seas. They feed here until late summer, when ice begins to force them out. In 1988, three Pacific gray whales lingered too long near Barrow, Alaska, and were imprisoned by advancing pack ice. The youngest whale died, but the heroic rescue of the other two, involving a community of Inuit, helicopters, bulldozers and, finally, a Russian icebreaker, was anxiously watched on televisions around the world.

As remarkable as the Pacific gray whale's migration is its near-miraculous return from the brink of extinction. Gray whales were essentially ignored by early nineteenth-century whalers: their baleen plates, used to make corsets, were too small; their oil was less plentiful and inferior to that of other whales; and the meat wasn't especially good. Unless they could be taken by the hundreds they weren't worth the effort. But with the discovery of their Mexican breeding lagoons in the mid-1800s, gray whales could be easily slaughtered in great numbers.

In the lagoons, the gray whale earned the nicknames "hard head" and "devil-fish" because of the way enraged females aggressively defended their young. In spite of the confrontations, the whalers were the ultimate victors: whale sightings in the lagoons soon became a rarity. The ships abandoned the lagoons, leaving only shore-based whalers who set up gauntlets of small boats on the migration route.

According to Captain Charles Scammon, a whaler who wrote a book on marine mammals of the northwest coast, other, less sophisticated whalers were also familiar with the gray whale's migratory habits. "Scarcely have the poor creatures quitted their southern homes before they are surprised by the Indians about the Strait of Juan de Fuca, Vancouver and Queen Charlotte's Islands," Scammon wrote. "Like enemies in ambush, these glide in canoes from island, bluff, or bay, rushing upon their prey with whoop and yell, launching their instruments of torture, and like hounds, worrying the last lifeblood from their vitals."

Whale hunts were immersed in ritual, and log books from early trading vessels suggest the Indians of Ahousat, near Long Beach, customarily sacrificed a slave in honor of the first whale killed in a season.

They hunted from canoes, using harpoons with sealskin floats attached by long lines of sinew and cedar bark. The whale, taking refuge in the depths, would fight against the buoyancy until exhausted. A lance would then be stabbed into the heart and, with the help of the floats, several canoes would tow the carcass ashore.

There is disagreement among researchers about the whale species taken by west-coast natives. Some say gray whales were the most important; others believe they hunted humpbacks. Barnacles that grow on whales have provided some clues for historians: barnacles from humpback whales, not grays, have been unearthed more frequently from Indian middens near Ahousat.

Whether they took humpbacks or grays, the effect of Indian whaling on the overall Pacific gray whale population was insignificant compared to the devastation by commercial whalers. By the time Scammon's book was published in 1874, the species was in serious trouble. "The mammoth bones of the California gray lie bleaching on the shores of those silvery waters, and are scattered along the broken coasts, from Siberia to the Gulf of California," he wrote, "and ere long it may be questioned whether this mammal will not be numbered among the extinct species of the Pacific."

With perhaps only two thousand gray whales left, they were virtually forgotten. Their numbers grew, unmolested, until they were rediscovered around 1914, when modern factory ships sailed into the scene. The First World War stifled some of the whaling, but soon after the war the butchery was revived with vigor. It wasn't until 1946, with the formation of the International Whaling Commission, that the Pacific gray whale was given long-overdue protection.

Today researchers are looking at the potential effects of whale watching.

BELOW: PACIFIC GRAY WHALE DIVES OFF WESTERN VANCOUVER ISLAND.

JIM DARLING

As competition increases, there is concern that whales will suffer undue stress by charter-boat operators eager to satisfy their customers. So far there's little to indicate Pacific grays are disgruntled with human intrusions. Some whales, in fact, appear to enjoy being the center of attention. Whale watchers are often greeted by "friendlies," whales that snuggle up to tour boats, as if enjoying the affectionate touch of a human hand.

We are fortunate these whales were not lost forever. Their recovery from near-extinction to historic numbers is unparalleled among the great whales. Today, as in the last century, the Pacific gray whale supports a thriving industry. This time, however, the industry threatens neither its freedom nor survival as a species.

Minuscule Meals for Massive Mammals

Feeding gray whales can often be located by the trails of sand and silt they churn up as they scrape along the seafloor, gobbling crustaceans, tubeworms, amphipods, and other invertebrates. By devouring tonnes of these minute marine animals, in eight or ten years a gray whale may grow to thirty-one tonnes, measuring thirteen metres from its barnacle-encrusted snout to its flukes. No one is sure how long a gray whale lives, but forty or fifty years is not unlikely.

Like other baleen whales, Pacific grays filter out tiny prey through a series of plates which grow down from the roof of the mouth. Similar in chemical composition to the human fingernail, the outer edges of the baleen are smooth, while the inner edges are lined with bristles or fibers. The whale sucks in huge quantities of sand, water, and food, then expels it through the baleen plates with the force of a fourteen-hundred-kilogram tongue. The bristles on the inner edges of the baleen catch tiny organisms as the water and sand shoot out. Scientists have discovered consistently greater wear of baleen plates and an almost total lack of barnacles on the right side of gray whales, suggesting they swim on that side while feeding.

Humpback Whales — Deep-Sea Singers

Humpback whales are the singers of the sea, known for their eerie, high-pitched vocalizations that reverberate through the depths. Apparently only males sing, repeating musical phrases for half an hour at a time. There appear to be different songs for different populations and scientists believe the use of their wide "vocabularies" may be an expression of dominance or aggression.

Like other cetaceans, humpbacks were hunted relentlessly. Today, with perhaps only five thousand left, some whale experts believe the North Pacific humpback's place in the twenty-first century remains precarious. Near their summer feeding waters, between southern California and the Bering Sea, they are imperiled by fishermen's nets, oil exploration and production, habitat destruction, and pollution from ocean dumping. Near their winter homes in the tropical waters of Hawaii and Mexico, they face growing human disturbance from resort development and ever-increasing numbers of tour boats and cruise ships invading their calving grounds.

GRAEME ELLIS

BELOW: HUMPBACK WHALES STAGE SPECTACULAR ACROBATIC SHOWS.

Pacific Rim Whale Watching

The annual migration of gray whales is celebrated on Vancouver Island each spring with the Pacific Rim Whale Festival. Films and slide shows are shown in theaters at Long Beach, in Pacific Rim National Park. Guided "whale walks" are given by park naturalists, who take people to prominent viewpoints and explain how to spot surfacing whales.

Here a naturalist shows a diagram of a whale and tells his audience to scan the sea for the telltale spouts beyond the surf line. After a whale dives it reappears farther north. Whale watching requires patience, and many observers at Long Beach are rewarded by spectacular displays of flashing flukes and breaching whales.

As the migration peaks, around March or April, literally thousands of tourists take to the sea aboard whale-watching boats based in the villages of Tofino and Ucluelet, at opposite ends of Long Beach.

PREVIOUS PAGES: HUMPBACK WHALES ARE KNOWN FOR THEIR EERIE UNDERWATER SONGS.

(Graeme Ellis photo)

In their mating waters, these acrobatic whales entertain thousands of tourists with their fluke-flashing, breaching, and spyhopping. They leap clear out of the water, twisting their titanic bodies in half-turns to crash down on their sides in frenzies of foam.

As these whales breach, they are particularly easy to identify by their long, winglike foreflippers with scalloped edges. The flippers of a full-grown humpback may measure five meters long, about a third of its total body length. The serrated flukes may be five meters wide. The tops of their heads are spotted with bumps resembling gigantic warts, each embellished by a single hair. Their skin is scarred, and barnacles cling to their chins, throats and flippers. Deep grooves in their throats stretch down to their navels. About two-thirds of the way back from the head is a tiny dorsal fin, sitting on top of a small hump. Humpbacks often lie on the surface with a flipper in the air, occasionally using it to communicate by slapping the water.

Males also slap one another with their flippers when competing for females. Couples mate on the calving grounds between December and April. Calves, measuring almost five meters long and weighing two tonnes, are born here and nursed by extremely protective mothers.

At the end of the calving season, humpbacks leave their breeding waters and head for summer feeding grounds, relying on their instinctive navigational skills as they journey thousands of kilometers across the open sea.

Small numbers come to the Pacific Northwest, in May or June, and forage offshore, beyond sight of most whale watchers. Occasionally a humpback swims through Juan de Fuca and Georgia straits, or down into Puget Sound. They have been seen in large sounds and in the deep, steep-sided fjords of B.C. and Alaska.

Researchers have identified about three thousand individual humpbacks off North America's west coast and new whales are being found constantly. Flukes, dorsal fins, and skin pigmentation carry markings that clearly distinguish one whale from another. The ability to identify individuals has helped scientists understand their behavior and migration habits. In 1980 a humpback seen feeding off Swiftsure Bank, near the entrance to Juan de Fuca Strait, was also seen at Hawaii in the winters of 1980 and '81. The same whale was seen again at Swiftsure Bank in 1989.

In the cool, productive, summer feeding waters, humpbacks eat krill, herring, sardines, and other small fish, sifting food through baleen plates that hang seventy-five centimeters from their upper jaws. Sometimes they lie leisurely on their sides, sucking in prey on the surface. A more aggressive hunting method is known as "lunge feeding": they plow through the water with their mouths agape, scooping food as they swim. Their most intriguing technique is "bubble net feeding." Below the surface, one or two whales swim in circles around a school of fish, blowing bubbles as they move. A mass of bubbles rises toward the fish like a circular net. The frightened fish gather in a tight ball in the center of the net and the whales swim up from below with their mouths wide open.

In the days of whaling and whale oil a full-grown thirty-five-tonne humpback could yield 40 or 50 barrels of oil. One humpback taken off California in 1858 yielded 145 barrels of oil. Whalers are no longer a worry, and a humpback who survives modern-day hazards may live almost as long as a person.

Minkes and Other Finners

The ten-meter-long minke, a year-round resident of the Pacific Northwest, is one of the smallest baleen whales. Oddly enough, it's among the whales known as finners, which include the twenty-six-meter, hundred-tonne blue whale, the largest animal on earth. The blue whale and two other finners that frequent the northwest — the sei and fin — are rarely seen by landlubbers or boaters who stay close to shore. They tend to forage offshore, far beyond the horizon. Little is known about these whales, compared to other species.

Minke whales are migratory and many spend their winters in southern

BELOW: THE MINKE IS ONE OF

THE SMALLEST FINNER WHALES.

climes. In the Pacific Northwest they are most often seen during summer, when they fearlessly approach boaters in Puget Sound and in the San Juan and Gulf islands. They are rather secretive and may surface two or three times near a boat, then vanish. They are also loners, though sometimes they travel in pairs or trios. Like gray and killer whales, minke whales occasionally wash ashore on Pacific Northwest beaches.

The minke is the most acrobatic of the finners, often breaching when feeding, jumping right out of the water, turning and coming down on its back. Like other finner whales, the minke feeds on fish and zooplankton and may live as long as a human being.

The seldom seen sei and fin whales are also migratory and are the fastest baleen whales, attaining speeds of twenty knots or more. They are much larger than minkes: an adult sei whale measures fifteen meters and a large fin whale may be twenty-four meters, almost as long as a blue whale.

Because of its small size, the minke was one of the last great whales to

GRAEME ELLIS

44

suffer at the hands of early whalers. It wasn't until the late 1800s, when bow-head and right whales showed noticeable declines, that more attention was focused on finner whales. The blue, of course, was the prime target, but the fin whale soon fell victim to the dreaded harpoon guns. As fin whales diminished, sei whales were next on the list, followed by minkes. Because finners haven't been studied as extensively as other whales, the numbers of these species in the North Pacific today are not well known.

Killer Whales — Sensitive and Sociable

Killer whales, with their proficient predatory skills, ominous, daggerlike dorsal fins, and the militaristic appearance of their traveling squadrons, have probably suffered more indignities in modern times than any other marine mammals of the Pacific Northwest. Known by some observers as "sea wolves," they

ABOVE: THE KILLER WHALE'S DORSAL FIN MAY STAND AS TALL AS A HUMAN.

have been targets for practising Canadian Air Force bombers; they have been shot by vandals and fishermen; they have been sentenced to lives of imprisonment in oceanaria around the world.

As far back as 1874, when whaling captain Charles Scammon published a book on marine mammals, killer whales, or orcas, were denigrated carnivores. "Indeed, they may be regarded as marine beasts, that roam over every ocean; entering bays and lagoons, where they spread terror and death among the mammoth balaenas and the small species of dolphins," Scammon wrote. "In whatever quarter of the world the Orcas are found, they seem always intent upon seeking something to destroy or devour."

Today our attitudes toward killer whales — undoubtedly the most spectacular of all marine mammals — have come full circle since Scammon's day. They have finally won long-overdue respect for their intelligence and unique social organization. No longer are they viewed as bloodthirsty killers of the deep, but as sensitive, affable beings. Ironically, or perhaps unfortunately, the ability to view captive killer whales in aquariums may be largely responsible for the turnaround in public sentiment toward the species.

Between 1962 and 1973, representatives of aquariums captured about 250 live killer whales from Pacific Northwest waters. Most were released; 11 died during capture operations or shortly after; 53 were taken to aquariums. One-quarter of the animals examined by researchers had bullet wounds.

Unprovoked attacks by killer whales on humans are rare: the story of a California surfer who survived a killer whale bite in 1972 is one of the few authenticated reports. Wild killer whales have attacked humans in self defence, however. In 1962, a pair of killer whales off Bellingham, Washington, charged a boat manned by whale catchers for Marineland of the Pacific. The crew had lassoed the female and the rope had become wrapped around the propeller. When the lasooed whale and accompanying bull charged the boat, the female was killed with ten shots from a rifle. One bullet was fired at the bull, which then vanished.

Attacks by captive whales are more common. Killer whales in aquariums have bitten or held trainers underwater, causing serious injury. The worst incident in recent history occurred in 1991, when a young trainer at Victoria's Sealand of the Pacific was drowned by three killer whales. The woman had accidentally slipped into the whale pool and was climbing out when one of the whales pulled her back into the water. She was tossed about by the whales before she died.

Four days before this tragedy, Hyak, a twenty-five-year-old whale at the Vancouver Aquarium died. The two incidents heightened a long-simmering debate over the incarceration of killer whales for public display. Captive whales have provided opportunities for close study by scientists, and have given millions of people a chance to view live whales. Some people believe the "sacrifice" of a few whales for aquariums is worth the benefits. Others, however, say we have learned all we can from captive whales: scientifically there is nothing to be gained by confining them. It is questionable whether the whale shows staged at aquariums create a true impression of the behavior of wild whales. Jumping through hoops and singing on cue are not normal habits for wild whales. These shows, especially in California, are particularly humiliating in some aquariums, where killer whales are often ridiculously adorned with silly hats and giant sunglasses. Their trainers ride them like cowboys while audiences cheer like patrons at rodeos. One wonders if it's the whales, or the trainers, who are on display.

Though no orcas have been taken from the northwest for aquariums since the 1970s, efforts to catch them elsewhere are invariably met with strong, often emotional, opposition. Activists have thwarted capture operations and, in 1982, an attempt to release a captive whale backfired. Miracle, a killer whale at Sealand, drowned while trying to swim through a hole which apparently had been cut in its mesh enclosure by a scuba diver. The whale had been rescued from certain death in 1977 when it was found near Duncan Bay, on eastern Vancouver Island, dying of gunshot wounds and malnutrition.

Scientists have documented about 390 killer whales that regularly patrol the waters of the Pacific Northwest, swimming in pods, or close-knit families, that stay together for life. These populations are growing by about 3 percent a year, perhaps in response to the captures of the 1960s and '70s, and possibly because fewer people shoot them nowadays. Historic numbers of killer whales are unknown, and only time will tell how many orcas the waters of the northwest can sustain. In recent years, new pods have been discovered offshore, and it's anybody's guess exactly how many killer whales roam the open seas of the North Pacific.

The orcas of the Pacific Northwest have been studied extensively since the 1970s by researchers who have pioneered techniques of identifying individual whales through photographs of dorsal fins, nicks and scratches, and skin colorations. Killer whales emit whistles and other sounds for communication, and clicks for echolocation. Recordings of their vocalizations indicate that each pod has its own dialect, a discovery that provides another way of identifying specific pods in the wild.

Pods may have as many as fifty members, but groups of five to twenty are more common. In the Pacific Northwest, pods join one another to form distinct communities. One community, with three pods totaling 89 whales, inhabits the waters of southern Georgia Strait, Puget Sound, western Vancouver Island, and the west coast of Washington down to Grays Harbor. A second community, comprising sixteen pods with a total of 190 whales, cruises between northern Vancouver Island and Prince Rupert. The two groups seem to respect an undrawn boundary in northern Georgia Strait: each stays on its respective side. These are "resident" whales, animals which are frequently seen in summer.

A third community is made up of about a hundred whales from forty-five small groups that move throughout B.C., Washington, and southeast Alaska. Because of their elusiveness they are known as "transients," and they don't mingle with resident whales.

With the help of fishermen, lightkeepers, mariners, and others who work and live on the coast, researchers have been able to track Pacific Northwest killer whales and study their behavior. One significant, and unexplained, difference between transient and resident whales is that transients are the fearsome foes of marine mammals, surviving on seals, sea lions, dolphins, and whales much larger than themselves. Residents subsist mainly on salmon, squid, octopi, and other fish. Resident pods are larger and have different traveling habits than transients. Residents apparently are more vocal than transients. There are even subtle physical differences: while transients generally have pointed dorsal fins, residents usually have dorsal fins which are more rounded at the tip.

In the Pacific Northwest, killer whales are seen most often in summer and fall. They are a suspenseful sight, as they break the surface, willowy clouds erupting from their blowholes. The dorsal fins of full-grown bulls stand as tall as humans, slicing through the water as they porpoise along at speeds up to fifty kilometers an hour. Wild killer whales sometimes amuse boaters and ferry passengers with their breaching, spyhopping, and tail slapping. As they spiral clear out of the water, they are easily recognized by their jet-black backs and white undersides, saddle, and eye patches.

Orcas are the largest members of the dolphin family, with nine- or ten-tonne bulls measuring almost ten meters long. Females of five or six tonnes may be more than eight meters from nose to tail. Sustaining a pod of killer whales requires tonnes of food: each adult may consume seventy-five kilograms of seafood a day.

Female killer whales may live an average of fifty years in the wild, while males may average thirty. It is not unknown for a cow to reach eighty or ninety years old, or a male to live to sixty. Killer whales, however, are slow to reproduce: a sexually mature male would likely be twenty-one years old. Females don't usually give birth to their first calf until the age of fifteen. A cow may produce five or six offspring and stop reproducing at about forty years old.

In spite of their apparent scarcity, killer whales are the basis of a flourishing tourist industry. In Washington, tour boats run out of Grays Harbor and the San Juan Islands. In 1984 a park was established at Lime Kiln Point, on San Juan Island, and a viewing tower was erected for whale watching. Orcas often pass close to shore off the point. The waterfront off Victoria, not far from the San Juans, is also fairly reliable whale-watching territory. Vancouver Island's most popular place for killer whale observation is Robson Bight, in Johnstone Strait, where whales stop to rub on gravel beaches. Tens of thousands of sightseers in charter boats, cruise ships, private boats, kayaks, and canoes come to see the whales. In 1982 the British Columbia government declared Robson Bight an ecological reserve to ensure protection of the whales' habitat. In 1991 the reserve was renamed Robson Bight-Michael Bigg ecological reserve. Dr. Bigg, who in 1990 died of cancer at the age of fifty, had studied Pacific Northwest killer whales for two decades and was considered by many to be the world's foremost authority on the species. He developed the system for whale identification and inspired numerous researchers and volunteers to help study wild killer whales.

BELOW: BREACHING KILLER WHALE, LARGEST MEMBER OF THE DOLPHIN FAMILY.

FRED FELLEMAN

Wolves of the Sea

The killer whale received its rather unflattering common name because it is the largest predator in the world to eat warm-blooded animals. Found in every ocean on earth, they are most plentiful in colder seas. Some populations feed exclusively on fish, while others prefer seals, sea lions, smaller cetaceans, penguins, or baleen whales much larger than themselves.

Often referred to as "sea wolves," killer whales that feed on marine mammals may hunt in packs. Seals or sea lions are playfully tossed about for an hour or longer before being devoured. These whales have large throats which allow them to swallow other mammals with little trouble. Packs of killer whales feed on larger whales by tearing off large chunks of flesh.

Only in recent years have these magnificent cetaceans been viewed as creatures to protect and admire. During the 1950s, many countries considered the killer whale vermin. The U.S. Navy reportedly machine-gunned hundreds in the North Atlantic at the request of the government of Iceland.

Dolphins and Porpoises — The Little Whales

Even among seasoned sailors there is still confusion over dolphins and porpoises. All fall within the suborder Odontoceti, or toothed whales. To take it one step further, dolphins are members of the family Delphinidae and in the North Pacific there are nine species in this family, including the killer whale. Three North Pacific species comprise the family Phoceonidae, or porpoises. These terms, however, are technicalities and the words "dolphin" and "porpoise" are often interchanged in conversations around wharves and fishboats.

Around northwest waters, one of the most visible porpoises is the Dall's, one of the fastest cetaceans, capable of speeds up to fifty-five kilometers an hour. These year-round residents are the most common small cetacean. They are bow riders, but easily become impatient with boats that travel under sixteen or eighteen kilometers an hour. They are most often seen rolling forward on the surface, several rising simultaneously as they swim. They are difficult to photograph because they rarely jump out of the water.

With its rotund body, triangular dorsal fin, and black and white colors, the Dall's porpoise is occasionally mistaken for its larger kin, the killer whale. This little porpoise, however, is two meters long and weighs about two hundred kilograms, rather puny beside a ten-thousand-kilogram, nine-meter-long killer whale.

Dall's porpoises and harbor porpoises are often seen together in the mainland fjords of B.C., in the Queen Charlottes, Gulf and San Juan islands, in Puget Sound, and far out at sea. Pacific white-sided and northern right-whale dolphins also travel together. Most of their time is spent at sea over the continental shelf, but they do travel into Juan de Fuca and Georgia straits. Pacific white-sided dolphins, which swim into Queen Charlotte Strait in fall and winter, are a gregarious breed, traveling in herds up to a thousand. False killer whales, short-finned pilot whales, and Risso's dolphins are other members of the dolphin family that periodically journey through the Pacific Northwest.

OPPOSITE: DALL'S PORPOISES OFTEN TRAVEL WITH HARBOR PORPOISES.

BELOW: PACIFIC WHITE-SIDED DOLPHINS ARE OCCASIONALLY SEEN IN PUGET SOUND AND JUAN DE FUCA AND GEORGIA STRAITS.

F. STUART WESTMORLAND

Land Mammals of the Coast

THE SHORES AND WOODLANDS ABOVE Pacific Northwest seas are home to an array of wildlife. While the large mammal species here are found inland, many are also coastal animals. The tracks of bears, cougars, and deer are a familiar sight on remote beaches; wolves and racoons have learned to probe the intertidal zones for the resources of the sea.

The land and sea are inseparable. Rivers pour down from the mountains; where they meet the sea they form estuaries, the most productive environments on earth. These vital habitats for waterfowl, migrating fish, raptors, shorebirds, and more are the link between land and sea: the effects of our industrial endeavors far offshore, or far upstream, can be judged by the health of our estuaries. If we fish with more avarice than sense, the salmon will not return, affecting bears and birds upstream. If we harm our mountain streams, the fish that feed whales and sea lions will have nowhere to spawn. Though the land and sea are entirely different environments, the relationship between marine and terrestrial wildlife is far from remote.

Pacific Northwest Bears — Grizzlies, Blacks, Kermodes

In both ancient and modern times, the bear has been viewed with mixed emotions. In old-world fables and Indian legends it has symbolized evil and aggression, power and valor. It has been feared for its great strength, exalted for its courage, and nervously respected for its unpredictability.

Only since the advent of television, with the animal harems of Grizzly Adams and the Hollywood-born beasts of Disney's World, has the bear been portrayed as a cuddly, endearing honey-lover that chases butterflies and yearns for human companionship. On the contrary, bears are reclusive, avoiding human contact as they forage from alpine tundra to valley bottoms. Sadly, those that do encounter people are often enticed by garbage dumps, campgrounds, or tourists with handouts: these unfortunate creatures then are categorized as "problem wildlife," vermin to be shot for the hazards they impose upon humans who trespass in the bear's domain.

Gradually, as our understanding of wildlife habits and habitats improves, our attitudes are changing. The bear is more readily viewed as an integral part of the natural balance, as an animal to be protected and admired. We're admitting that humans, not animals, are the creators of "problem wildlife" and we're cleaning up our garbage dumps, camps, and outback communities.

Yet despite this apparent headway we continue to stymie the bear's freedom. Habitats vanish at a shameful pace; hunters hang new trophies on their walls; poachers ship bears in bits and pieces to the Far East. Although there is no shortage of bears in the Pacific Northwest today, there is also no guarantee they will be here forever.

Grizzlies — Caught in a Land Squeeze

The Linnaean name *Ursus arctos horribilis* leaves little room for speculation about the grizzly bear's status. It is among those creatures to which humans have attached fears of mythic proportions, ranking with sharks and crocodiles as a beast to avoid. Oddly, the name "grizzly" is not a reference to its disposition, but an allusion to its coarse fur.

A big grizzly boar may weigh four hundred kilograms in its prime and stand upright at more than three meters. When lumbering along on massive forelegs, the top of its distinctive hump may be more than a meter from the ground. One forepaw, which could measure twenty centimeters across the palm, is well over twice the size of a human hand. With long, curved claws, it uproots

small trees or obliterates the burrows of badgers, rodents, and other prey.

Grizzlies sprint as fast as race horses and use short bursts of speed to take newborn or young deer, elk, and cattle. The common perception of grizzlies as ferocious meat eaters, however, is misleading: they are omnivores, surviving more on plants than on fish and carrion.

Big bears need big territories: an adult boar may search an area of 350 or 500 square kilometers to sustain itself year-round. A sow may cover 100 or 150 square kilometers. Foraging areas may overlap and much of their feeding is in valley bottoms and estuaries.

On the Pacific coast, grizzlies emerge from their dens in late March or April. Sows with cubs may sleep until May or early June. In spring they wander the valley bottoms, eating sedges in meadows and marshes, or skunk cabbage in lowland bogs. Berries and succulent new growth in avalanche chutes and other areas sustain them through summer until autumn, when salmon begin to appear in estuaries and streams.

With spawning done, the bears, fat and fit, move above valley bottoms and crawl under uprooted trees, or into caves or underground dens, and drift into a deep torpor. They are not true hibernators and are known to awaken in midwinter, take a short stroll, then sleep another month or two.

The slow reproduction of grizzlies makes them vulnerable to excessive hunting or habitat losses. A mature, six- or seven-year-old sow will usually produce only two cubs every third year. With most bears living less than twenty-five years, it could take many "bear generations" to successfully restock a watershed from which a population has vanished.

ABOVE: KHUTZEYMATEEN WATERSHED IN NORTHERN B.C., SITE OF EXTENSIVE GRIZZLY BEAR STUDY.

Grizzlies once ranged as far south as Mexico, but now there are few places where they are part of the natural fauna. Oregon has none. Washington shares a small population with Idaho, as well as a few with B.C. in the Cascade Mountains. Alaska still has healthy numbers. British Columbia's total is estimated at more than twelve thousand, with about thirty-five hundred roaming coastal watersheds.

Grizzlies can survive in alpine tundra or second-growth forests, but mature forests, with their nutritious understory, are a bear's best habitat. These primeval woodlands are self-sustaining, with trees of varying ages, from saplings to

The Grizzly Fisher

Spawning salmon in autumn are an important source of protein for grizzly bears, and the fishing season is the most likely time to see congregations of grizzlies, often two dozen or more. Groups of bears may fish together, though not cooperatively, occasionally quarreling when one oversteps another's chosen stretch of stream. Black bears, eagles, herons, gulls, ravens, mink, otters, and other animals may feed on dead salmon near fishing grizzlies, keeping a respectful distance.

Some grizzlies are skillful fishers; some use cruder methods. Fishing techniques vary, depending on conditions and currents. One method is to wade into the riffles and watch patiently for a fish to pass. At the right second, the grizzly dips its head underwater and snatches the fish in its jaws.

Sometimes a bear holds the fish against the streambed with a heavy paw, then picks it up in its teeth. In shallow streams a grizzly may herd a salmon into a channel or right up onto shore.

In spite of their size, grizzlies are fastidious eaters. A salmon is usually taken to a secluded place among the trees, where the bear delicately strips off the flesh, frequently leaving the skeleton intact. The gills and viscera are rarely consumed, and if fishing is favorable a grizzly may eat only the belly and cheek muscles.

centuries-old specimens. When an old tree falls, the forest opens, allowing sunlight and rain to nurture new growth. Second-growth forests are planted all at once, so trees don't grow at staggered rates. The canopy of a twenty-year-old second-growth forest is closed, blocking the sun and rain needed to raise a productive understory. What little food there is for a bear in these new forests is of inferior quality.

Yet in coastal B.C. the ancient evergreens are being liquidated at an alarming rate. If the current pace continues, within the next couple of decades the most valuable timber will likely be gone: true coastal wilderness may exist only in disjointed patches incapable of supporting viable populations of grizzlies or other wildlife.

Concern over habitat losses and the grizzly's slow reproduction have prompted resource managers to find ways of integrating sustainable forestry with grizzly-bear management. By the mid-1970s there was a noticeable decline in grizzlies occupying logged watersheds in coastal B.C. New roads harmed the environment and improved accessibility for hunters and poachers; open clearcuts made bears more visible; bears were shot in defence of timber cruisers, tree planters, or forestry crews.

Valley bottoms and lower slopes have the easiest access and hold the most valuable timber. They are also critical foraging areas for bears. Estuaries are salmon-feeding grounds; floodplains and avalanche chutes produce edible vegetation. But the clearcutting methods favored by the logging industry today injure sensitive spawning streams. Although replanted forests regenerate with little difficulty, some foresters prefer to speed the growth of new trees by spraying herbicides on the understory, which bears eat. Logging also lowers the water table in bogs, which support skunk cabbage, a staple in a grizzly's diet.

Logging destroys not only food sources, but other habitats. Bears may abandon foraging grounds if adjacent bedding areas vanish. They also sleep in forests next to avalanche chutes, which they use as travel corridors between watersheds.

Answers to industry-nature conflicts are a matter of compromise. With lumbering as B.C.'s economic base, the environment has traditionally been the loser in these conflicts. Today, as the last of our virgin forests are systematically clearcut, the public is skeptical of the forest industry's sustainability. More now than ever, the industry is being challenged to prove its "forests forever" slogan, to honestly show its willingness to compromise. True sustainability cannot be one-sided: it means the continued existence of environment, wildlife, and industry.

BELOW: GRIZZLY SOW NURSES HER CUBS.

Black Bears — Source of Dubious Tales

The future for black bears — *Ursus americanus* — is brighter than for grizzlies. They are smaller, requiring less food. They need about one-quarter the range of grizzlies. They are adaptable, more tolerant of invasions and changes in their environments. They mature sooner, reproduce more often, and exist in much greater numbers.

Anyone who explores the northwest outback is bound to encounter a black bear sooner or later. They meander onto beaches and raise their noses at passing boaters, depending on their powerful olfaction to compensate for mediocre eyesight. When they rear up on hind legs at intruders, they are more likely trying to get a better look than preparing to charge. They hang around unkempt campsites and dumps, scavenging garbage, the objectionable byproduct of many family meals. They wander up roads and riverbeds, chasing cubs up trees at signs of approaching danger.

Though exceedingly dangerous when riled, black bears don't look as terrifying as grizzlies, probably because of their smaller size. Large boars may weigh two hundred kilograms, half as much as grizzlies. Their claws are shorter and they lack the portentous hump that gives grizzlies such an aggressive posture, as if their backs are always up.

Black bears are nonetheless animals to avoid in the woods. The mere mention of the word "bear" around a campfire elicits tales of encounters with black bears, many true, some spiced with dubious detail. Like people, animals prefer paths of least resistance, so hiking trails are attractive to bears, particularly at night. Many attacks occur on trails, where bears, sometimes with cubs, are surprised. Bears are especially aggressive toward people wearing scented cosmetics, hair spray or deodorant, or women who are menstruating.

In the majority of human encounters with bears, the animal takes a quick sniff and dashes into the bush, a natural reaction from a frightened creature. But a less desirable reaction is to charge, knocking the victim to the ground. Outdoors experience can't help a person in that predicament. Many attacks, however, are provoked by appalling stupidity: shutter-happy parents have been known to place children on the shoulders of park bears, or to use food to lure bears into family vehicles.

In these situations, which have brought tragic results for humans, the victims are the bears. Like bears attracted to improperly managed garbage dumps, they have lost their fear of people, and therefore pose a threat. Trapping and transplanting a "problem bear" is usually futile. When introduced to a new area, it upsets an established order and is chased out, or killed, by existing inhabitants. A survivor returns to its old haunts, where daily meals are delivered on schedule. If the handouts stop, the bear becomes aggressive, leaving little choice but to destroy it. Some bear biologists are trying to eliminate the phrase "problem bear" from our wildlife lexicon to raise the respectability of black bears. Their value, like that of grizzlies, may one day be their scarcity.

Hunting, and the serious offence of poaching, also takes its toll on bears. In B.C. alone, hunters legally kill four or five thousand black bears a year. Some wildlife authorities speculate that the illegal kill by poachers could be as much as 30 or 50 percent of the legal harvest. As populations of Asian bears teeter on the verge of extinction, the illicit trade in North American bear parts appears to increase. Gallbladders become Asian medicine; paws are a delicacy. In other parts of the world, black-bear hides hang on the walls of mansions; claws, femurs, and other bones are carved into jewelry.

Vanishing wilderness, as with grizzlies, is

LEFT: BLACK BEARS ARE
ABUNDANT ALONG THE PACIFIC
NORTHWEST COAST.

a worrisome factor in the endurance of the black-bear species. Today they are plentiful, but so were grizzlies only a few decades ago. In western Washington the value of bears as an integral part of the forest ecosystem is well recognized. Foresters and biologists have devised a way to spare the lives of bears that destroy second-growth forests. In spring, when the cambium, or growing layer, on young trees is formed into sapwood, the bark strips easily. Black bears tear off the bark and gnaw on the exposed cambium: its sugar is an important nutrient to a bear that has just emerged from a winter-long snooze. They eat the sapwood until summer, when the berries are ripe. Unfortunately, this particular eating habit destroys the forest. For the past forty years the solution was to shoot the bears.

Today's answer is to give the bears something better to eat — supplemental food pellets that reduce the bears' preference for trees. In 1986 nearly sixteen hundred kilograms of pellets were eaten at 9 feeding stations in Washington's coastal woodlands. The next year bears ate twenty-five hundred kilos at 10 stations. It was obvious this plan was working, so in 1988, twenty tonnes of feed were set out at 160 stations spread over forty-two hundred square kilometers. In 1987, 69 percent of the trees in test plots showed improvement: the next year it was 85 percent. Owners of tree farms began using the feeders, and for nominal investments their future marketable timber remains intact.

Kermode — Snow Bear of the Northwest Coast

The Tsimshian Indians say there is a ghost that stalks the coastal rain forests of northwestern British Columbia, fighting evil, rescuing people in distress and diving deep into the sea to hunt seals, whales, and halibut. This ghost, they say, is not a human spirit but a white bear with power to take human form.

Today this elusive beast is known by some as the snow bear, a pure white wanderer that forages for bugs and berries, hunts spawning salmon, and crawls into a hole to sleep through the cold northern winter. Science books call it *Ursus americanus kermodei*: most people just call it kermode.

The white kermode (pronounced ker-mode-ee) is, in fact, a black bear. The term "black bear," to add confusion, denotes a bear species, not a color. While most black bears are actually black, they also come in various shades of brown, from blonde to cinnamon and auburn. One is the kermode, a black bear of many hues that inhabits a limited range bounded roughly by the communities of Stewart, Prince Rupert, Hazelton, Terrace, and Kitimat. It appears to be most plentiful along the rugged shores of Douglas Channel and on Gribbell and Princess Royal islands, south of Kitimat. When tourists ask where they can see a kermode, they are invariably directed to the Terrace dump.

While it is the white bears that are the mythical beasts of Indian folklore, kermodes may also be chestnut red, orange, yellow, bluish-gray, gold, black, brown, or a mix of black, brown and white. It's not unusual for a white kermode to produce black offspring, or vice versa. Though not the most common, it is the white kermodes that have attracted the most attention. They have, at times, been mistaken for misplaced polar bears. It once was believed they were a race of albino bears, but their brown eyes dispelled that theory.

There also have been disputes over its status as a separate bear species, or subspecies of the black bear. *Ursus kermodei* was originally designated a

distinct species by Dr. William Hornaday, a naturalist with the New York Zoological Society. In the 1905 annual report of the society, Hornaday not only described the unique characteristics of his newfound species, but he portrayed northern B.C. as a land virtually undiscovered by the scientific community: "There are vast regions, containing we know not what new animal life, which have been practically untouched by the zoologist. . . northern British Columbia is, to scientific collectors and students, a land almost unknown."

Five years earlier, while cataloguing bearskins in Victoria, B.C., Hornaday ran across a creamy white pelt that had come from the Nass and Skeena river territory. He asked Francis Kermode, director of the B.C. Provincial Museum, to investigate the existence of the white bear. Kermode had never heard of these ghostly creatures and, with few inhabitants in northwest B.C., sightings were scarce. In 1904, however, Kermode acquired three more white pelts, all from the same region. They were given to Hornaday, who declared "a new species of bear," which he named in honor of his colleague.

Hornaday's declaration was challenged in 1928 by E. Raymond Hall, a mammologist at the University of California in Berkley. While Hall agreed that the kermode had cranial and dental differences from the common black bear, they were not distinctive enough to warrant separate-species status. Its peculiar color variations, however, were sufficient to designate the kermode a subspecies.

As communities in the northwest developed, sightings of white kermodes, particularly at garbage dumps, began to increase. Word spread and there was growing concern about the need to protect this unique subspecies. In a province where hunters shoot an average of four thousand black bears a year, a white kermode would be a highly prized addition to a big-game hunter's trophy collection.

Hunting of white kermodes is banned, but because kermodes of other colors can produce white offspring, there's a strong case for protection of all black bears in the kermode's territory. In any event, a kermode born white invariably gets better treatment than its black cousin. So-called "problem bears" that invade campgrounds, garbage dumps, and remote communities usually meet untimely deaths at the hands of conservation officers. White kermodes, however, almost always get a second chance.

One kermode caught robbing lunches from parked trucks saved itself by crawling into a hole and sleeping for the winter just as conservation officers were considering capturing it. The bear had become accustomed to people because a photographer had been using fish to lure it to a convenient posing spot. The photographer had unwittingly baited a place where anglers parked their vehicles.

The kermode has been adopted as the emblem for the city of Terrace, where the white bear appears on the official flag, on stationery, T-shirts, and souvenirs. Even the municipal council chambers are embellished by a white kermode. Hank, the victim of a car accident, was stuffed and sentenced to eternally endure the ramblings of local politicians.

Surprisingly little research has been done on the kermode. No one knows the extent of its range, the size of the population, or if, in fact, it exists only in B.C. We do know, through the legends of the Tsimshian Indians, that the kermode bear has been here much longer than modern man. Yet it is modern man who will likely determine how long the kermode will stay.

Cougars — Pacific Northwest's Largest Wildcats

Few animals can match the predatory skills of the cougar. It carefully, silently, stalks its victim. Then suddenly, in two or three lightning-fast leaps, it springs onto the back of its prey and sinks its teeth into the base of the skull. If it's a clean kill, it's over in moments. If it misses, the big cat abandons the quarry and selects another.

A full-grown fifty-kilogram cougar can take an elk or moose five times its size, but deer are its primary prey. It will consume three-quarters of the carcass, making repeated visits over several days, covering the remains with dirt and debris after each meal. One adult cougar might take fifteen or twenty deer a year. Although cougars eat a variety of prey — mice, bear cubs, porcupines, mountain goats, and occasionally domestic stock or household pets — fluctuations in cougar numbers are closely tied to the availability of deer.

When deer are scarce, cougars may turn to livestock. Most attacks on domestic stock or people are by sick or old cats, but occasionally a healthy, mature male takes a farmer's sheep or cow. Some expert cougar observers theorize that these attacks stem from the death of a female in the area. A male cougar, like the African lion, is lazy, preferring to steal a kill from a female rather than hunt for itself. A tom will wander surrounding female territories, eating the kill of one, then moving on to the next female's area. If the tom arrives to find the female gone, along with the free lunch, it may kill a sheep, goat, or dog to replace that lost meal. Inquiries in these cases have revealed that often a cougar in the same area was killed shortly before the livestock or family pet was taken by the mature male.

Attacks on people and their animals are rare. Cougars are secretive, seldom seen by humans, even though they are often nearby. It is not uncommon for a cougar to perch high above a neighborhood or work site and watch the activities of people for hours, simply satisfying its feline curiosity. On Vancouver Island, which probably has the densest cougar population in North America, conservation officers remove one or two dozen cougars from residential areas each year. In downtown Victoria, in 1989, a cougar being pursued by wildlife officials with a pack of hounds was shot after it plunged through the window of an apartment. The outcry from animal lovers was deafening. Why wasn't it tranquilized and taken to some faraway forest? First, during the ten minutes needed for a tranquilizer to take effect, a panic-stricken cougar, with half a dozen frantic hounds on its heels, can inflict a lot of damage. Second, cougars are highly territorial, and with reasonably stable populations in the Pacific Northwest, few territories are unoccupied.

Cougars are loners, coming together only to mate. A single cougar needs forty or sixty square kilometers of hunting ground, which it repeatedly marks by urination and defecation. Where populations are healthy, the availability of territories may play a more significant role than the abundance of prey in keeping cougar populations in check.

Nowadays conservation officers make every effort to save a cougar. If there's no potential threat to human lives, and circumstances are favorable, a cougar will be drugged and, if possible, released in a territory that is known to be vacant.

For the first half of this century, however, cougars were vermin. In fact, the shoot-to-kill attitude was profitable. In B.C. a bounty of $7.50 was offered on

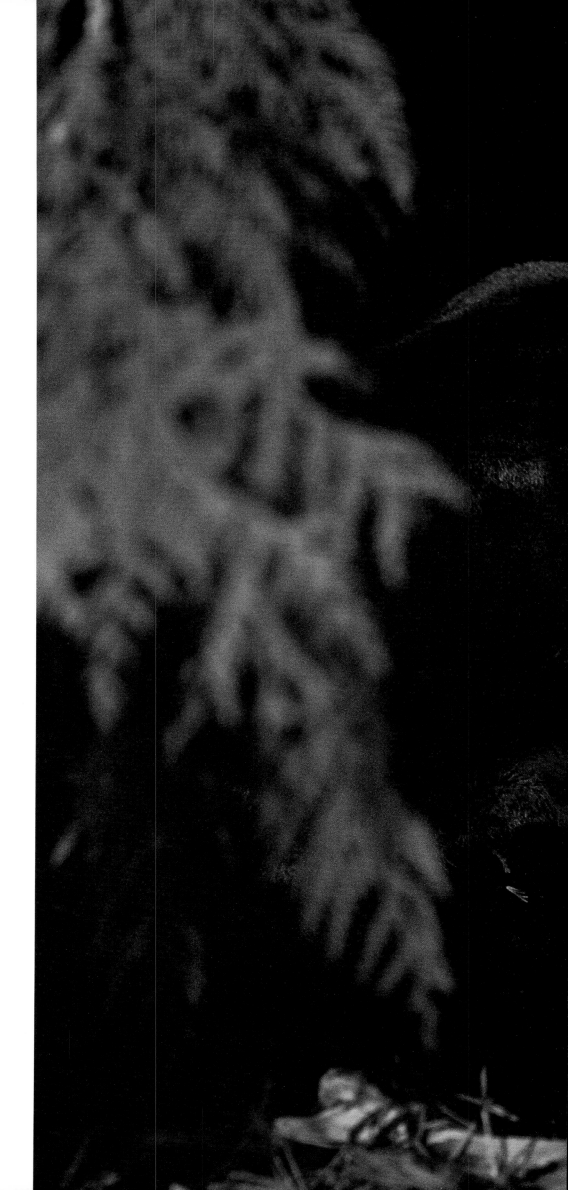

OPPOSITE: COUGARS OFTEN
WANDER THROUGH RESIDENTIAL
DISTRICTS.

cougars in 1906. By 1920 the government was employing hunters to destroy the species, and soon raised the bounty to $40.00. The amount varied until the bounty was discontinued in 1957. By then, 15,780 bounties had been paid. Today cougars are classed as big-game animals, which are taken by hunters who normally track them with hounds.

Wolves and Coyotes — Persecuted Canines

The somber howl of a wolf on a still autumn night is a symbol of true wilderness. It is a melancholy call, as distinctive as the laughter of the loon, a sound that lingers long in the memories of those who hear it. As intriguing as they are elusive, wolves arouse the curiosity of anyone who tramps the back-woods. A few people venture out specifically to find them, to imitate their calls and hope for a reply. Most are satisfied to catch an unexpected glimpse of a

BELOW: THE TRACKS OF WOLVES CAN BE SEEN ON REMOTE NORTHWEST BEACHES.

wolf dashing into the forest. Dog lovers occasionally attempt to domesticate wild wolves, but few are successful.

The orderly social structure of wolf families has intrigued wildlife observers for generations. Their interactions and noticeable affection for one another continue to fascinate us. Packs of five to eight individuals include a dominant breeding pair, pups of the year, yearlings and the occasional related adult. Normally the dominant pair breeds in late February or early March, to produce litters of four to seven pups. Born in May or June, the pups are blind for the first two or three weeks of their lives. They are cautiously guarded by the mother until weaned at about eight weeks.

During that time, the mother and young are fed by other pack members that regurgitate food at the den or carry it from kills. By autumn the pups begin to travel with the others, using territories of 130 to 1,500 square kilometers, depending on terrain, seasons, location, and abundance of prey. Young wolves dispersing from a pack could travel up to three hundred kilometers in search of a territory.

Wolves looking for food may move thirty or forty kilometers a day. In the Pacific Northwest the main prey is deer, but wolves also take beavers, rabbits, birds, elk, goats, and, occasionally, livestock. They leave their tracks on remote coastal beaches where they feed on fish and intertidal life.

ABOVE: COYOTES ARE PLENTIFUL IN B.C.'S FRASER VALLEY.

The hunting abilities of the wolf are legendary, and lie at the root of its reputation as a brutal, bloodthirsty killer. Hunts are preceded by a great deal of commotion: pack members howl repeatedly, and anxiously pace up and down. The key to their success is endurance — they can run for hours, tracking their prey until the luckless victim is too exhausted to continue, before closing in. A male wolf, weighing forty-five kilos and measuring two meters from its nose to the tip of its tail, can bring down a two hundred-kilogram elk with the help of the pack.

Misconceptions about the wolf and its hunting expertise have plagued the species since ancient times. In 1000 A.D. Europeans believed that wolves were enemies of rulers, and consequently belonged to the devil. Medieval illustrations depict Christ fighting wolves and show wolves dressed in human clothes, hanging from gallows — evidence of the werewolf myth. Man's historic fear of wolves is unfounded: wolf attacks on humans are extremely rare. Through early history, stories of wolves killing people surfaced most often during times of war and epidemic, which later led historians to believe that wolves were carrion-eaters, feeding on the already dead.

In contrast to this fear and loathing, the Indians of the northwest respected the wolf as a symbol of strength, courage, and wisdom. Bella Coola mothers painted a wolf's gallbladder on the backs of their children to ensure their spirituality and hunting capabilities. When Kwakiutl Indians killed a wolf for its skin, each of the hunters would eat strips of meat cut from the animal as an expression of regret at the wolf's death.

With European settlement, the wolf was persecuted throughout North America. They were long gone from Britain and parts of Europe in 1793, when the parliament of Upper Canada passed an Act to Encourage the Destruction of Wolves and Bears, an unequivocal statement on the status of predators in our society. By then Canadians were already 163 years behind Americans, who paid their first wolf bounty in 1630.

Wolves have since been wiped out of much of their traditional range by bounties, relentless hunting, deliberate poisoning, and habitat degradation. In the northwest today their numbers appear to fluctuate greatly.

The wily coyote, a relative of the wolf, is about one-third as large and looks like a cross between a wolf and a fox. Mating pairs may stay together through several litters, bearing five or seven pups a year, but they do not form lifelong relationships.

Coyotes, like wolves, have suffered at the hands of man, and large numbers have been taken for bounty and poisoned to protect farm animals. While they do occasionally prey on livestock or chickens, ironically coyotes can be beneficial to farmers because they subsist largely on pesky rodents. Coyotes have managed to expand their range and in the Pacific Northwest they are found in southwestern B.C. and northern Washington. They are particularly abundant in the Fraser Valley.

A coyote in the distance can be distinguished from a fox or young wolf by the way it carries its tail. Most canines raise their tails when on the run: a coyote's tail hangs low. Its most notable idiosyncrasy is its cry, a series of sharp yelps followed by a long, high-pitched howl. It is usually heard at dusk, and on old cowboy movies.

Blacktail Buckskin

Blacktail deer along the Pacific Northwest coast have been a valuable source of food and materials to humans for generations.

Soft and pliable buckskin, used for shoes, trousers, jackets, and other clothing, was originally tanned by Indians. Natives and early settlers often relied on tender, juicy venison for survival.

The blacktail deer's coarse fur once had a rather unusual use. The fur is stiff and tubular and during winter, when deer skins are particularly thick, they are also extremely buoyant.

These winter skins made good life preservers. Deer are strong swimmers and their dense fur helps keep them afloat as they travel across wide stretches of open water.

Blacktail Deer — Adaptable, Visible

Blacktail deer are the most plentiful big-game animals on the northwest coast. Unlike other species, which escape human encroachment by retreating farther into the wilderness, deer are adaptable. They are commonly seen on the outskirts of towns, in urban vegetable gardens, and on beaches. Timid but curious, they have a habit of quickly bounding away from danger, then stopping for a second look. In parks and well-worn hiking areas, many deer are accustomed to people and continue to forage unperturbed when humans pass close by. Blacktail deer are abundant on the San Juan and Gulf islands.

On the coast there are two subspecies of blacktail. Coast deer, or Columbian blacktail, inhabit the mainland forests, Vancouver Island, and offshore islands as far north as Rivers Inlet, about midway along the B.C. coast. Sitka deer, which are hardly distinguishable from coast deer, occupy the northern mainland, Queen Charlottes, and other islands into Alaska. Deer are strong swimmers and frequently cross wide channels to new foraging grounds.

During winter these animals are browsers, feeding on maple, salal, dogwood, western red cedar, Douglas fir, balsam, shrubs, and lichens. In spring and early summer, when grasses and forbs are high in protein, deer turn to grazing. Ideal deer habitat is a combination of forest with nearby patches of logged or open land. During winter in areas of heavy snow, deer seek the shelter of the forests, where dense cover reduces the amount of snow on the ground. They feed on nutritious undergrowth, lichens, and twigs which are blown to the forest floor. When the snow has melted they move onto open areas. On the islands of Georgia Strait and along the southeast side of Vancouver Island,

where snowfall is light, blacktail deer have little difficulty finding year-round food.

Each spring, when the blood level of testosterone rises, older bucks begin to grow new antlers. They become increasingly independent through summer, as they prepare for the autumn rut. By early fall the antlers are fully developed and the velvet which coats them is scraped off by rubbing against tree trunks and branches. During the rut that follows, the clash of antlers and grunts and snorts of quarreling bucks permeate the woods. Bucks that prove their territorial supremacy service the does and become more complacent as winter approaches. During winter bucks drop their antlers.

Does become sexually mature at about eighteen months and usually produce one fawn, occasionally two, which are born in late spring. Deer are known to live longer than twenty years, but in harsh natural conditions a decade is a long life.

Racoons — Downtown Wildlife

Racoons are downtown wildlife. Though they wander the wild woods and waterways of the Pacific Northwest, they are equally at home amid city streets and highrises, where gourmet meals are served in backyard fish ponds, composts, and trash cans.

With their pointed muzzles, dark masks and bushy, ringed tails, racoons are almost impossible to confuse with other animals. Coarse brown or silver hair covers all but the feet. They waddle with rumps in the air, as if crouching and walking at the same time.

The racoon's most enviable assets are prehensile forepaws, which it uses with a dexterity second only to monkeys. The hairless paws are as sensitive as human hands, and almost as useful. With five slender fingers it can undo chicken-coop latches, scoop goldfish from garden pools, even twist lids from jars. Sharp, curved claws rip open sweet corn husks, dig earthworms and other insects, crack birds' eggs, or dismember prey such as nestlings, frogs, mice, and squirrels.

One racoon tours a territory of two or three square kilometers. Suburban racoons follow established routes through their neighborhoods, traveling alone or in small groups under cover of darkness. On seashores, where tide pools teem with fish and crustaceans, racoons switch to daytime feeding to catch the low tides. Outside cities, they prefer forests and shrubbery near water, especially creeks with crayfish.

Although moderately sociable, racoons aren't especially gregarious. Monogamous females, which breed at a year old, are persnickety in selecting partners. Males, which don't mate until at least two years old or older, may breed with more than one female in a season. Males contribute little to rearing youngsters. Litters of one to seven kits are born in spring, and for a year the young learn the tricks and hazards of survival from their mothers. As adults, they measure a meter long and weigh eight kilograms. A ten-year-old racoon is old.

These animals don't hibernate but they do take extended naps during extremely cold weather. Adept climbers, racoons prefer to den in hollow trees about three meters above the ground. They also take up residence in crevices, caves, uprooted trees, culverts, or crawl spaces in houses. City coons are fond of attics, and roofs with cedar shakes are most accessible. They just rip off a

shake or two and move in, sometimes to give birth. Disgruntled homeowners who discourage racoons occasionally find their unwanted tenants are attracted by pet food left out by next-door neighbors.

The racoon's friendly face is deceptive, and people who entice them to their homes invite trouble. A racoon will aggressively defend a bowl of food or table scraps deliberately left out for it. If it arrives to find its usual meal not there, like a bear, it gets angry. People who approach might get a close look at the racoon's forty sharp teeth as it rears on hind legs and snarls. A cornered racoon is capable of disemboweling a dog three times its size.

When left alone, racoons are among our most entertaining wildlife. They are abundant on the northwest coast: chances are they are here for good.

75

Raptors, Ravens and Shorebirds

THE PACIFIC NORTHWEST'S MOST visible wild animals are the birds. Seabirds, shorebirds, divers, dabblers, raptors, and more — at least 250 species. Millions migrate through the coastal corridors of the Pacific flyway; hundreds of thousands nest in the rain forests, on the shores and islands; millions feed in the plankton-rich seas over the continental shelf.

Every bird, like every birder, has its idiosyncracies. The predatory powers of eagles, ospreys, and other raptors intrigue many wildlife watchers; the sheer numbers of migrating shorebirds draw people to beaches and coastal wetlands. The raven's trickery against other species, the nesting colonies and piscatorial aptitude of seabirds, or the daily pecking procedures of wintering swans continuously fascinate bird lovers. Even the much-maligned crow is not without its admirers.

Not only is watching birds one of the oldest pastimes, but it is one of North America's fastest-growing outdoor activities. The number of organized birders in the northwest has increased fivefold in a decade. It is estimated that 6 million American birders spend many millions of dollars a year pursuing their favorite avocation. Eighty-five percent of Canadians say they would happily invest more to watch birds and other wildlife. Some ornithologists claim that in the late 1980s *Peterson's Field Guide Series to North American Birds* outsold the bible.

Nonconsumptive use of our publicly owned wild animals is a growing trend, and increasing government attention to wildlife viewing reflects the change. The popularity of hunting and angling is still undeniable, but many shotguns and fishing rods are gathering dust in basements while their owners head out with cameras, binoculars, and spotting scopes. They are joined by a new breed of nature lover — once-casual afternoon hikers who now spend their weekend outings hunting down bird-watching hot spots. Armed with field guides, notebooks, and checklists, they are finding new dimensions to nearby environments.

The term "nonconsumptive," however, is debatable. As groups of wildlife watchers at seashores, wetlands, or nesting colonies proliferate, disturbances to birdlife increase. Birds, particularly when nesting, are easily stressed by human trespassers. Even simple observation has its impacts.

Fortunately, with the environmental concern of today a new conservation ethic is evolving. With common sense and caution, the pleasures of watching our avian animals can be enjoyed with minimal disturbance to the birds. Just as anglers and hunters contribute to the well-being of fish and wildlife, conservation-minded birders are becoming a major influence in the health and enhancement of our birds.

Birds of Prey — Avian Carnivores

Raptors are the inimitable predators of the sky. With their unparalleled aerial agility and formidable talons, they strike with a speed and force unmatched in the animal world. To watch one hunt is a rare privilege. Most astonishing is the peregrine falcon, which stoops at speeds over two hundred kilometers an hour, snatching its victim in a midair explosion of flying feathers. Some 140 bird species fall prey to the peregrine, the swiftest of all North American falcons.

To hunt successfully, birds of prey must judge distances with flawless precision. Unlike all other birds, whose eyes look in different directions from either side of the head, raptors have stereoscopic vision. Both eyes, comparatively much larger and more powerful than those of humans, face forward, allowing them to zero in on their prey with impeccable accuracy. Some species are able to focus on objects with a resolution eight times as sharp as that of people. This extraordinarily keen eyesight has evolved in raptors at the expense of good all-round vision. To see as broad a field as other birds, raptors must turn their heads almost full circle. This ability is most notable in owls, whose heads can arc as much as 270 degrees.

There is a certain nobility to birds of prey, and to see one soaring on the wind is an inspiration to any observer. Nearly three dozen raptor species, including hawks, falcons, harriers, eagles, owls, ospreys, and vultures, occur in the Pacific Northwest, many on the coast. The strigiformes, or owls, hunt day and night. The falconiformes, which hunt during the day, make up the majority of our birds of prey. On the coast, healthy populations of raptors are tied closely to prey: in areas such as the Queen Charlotte or Scott islands, with their plentiful seabird colonies, birds of prey thrive.

Good coastal raptor habitat consists of sheer cliffs, protected bays, open sea, forested islands, bare islands, mud flats, estuaries, rivers, and beaches. The intricate network of islands, tidal channels, and varied shorelines of Alaska, the Queen Charlottes, Vancouver Island, and mainland B.C. is prime raptor habitat. The long, continuous shorelines along the outer coasts of Washington and Oregon support fewer birds of prey.

These avian carnivores include not only the fastest, but some of the largest birds of the northwest. Among North American raptorial birds, the turkey vulture and bald eagle, with wingspans of two meters or more, are second in size only to California's endangered condor. With the exception of the turkey vulture, female raptors in the northwest are larger than their mates, usually by about one-third.

In the Pacific Northwest the bald eagle, with its snow-white head and tail, is certainly the most conspicuous bird of prey. Most of the world's bald eagles live here, mainly on the coast. Alaska has about thirty-five thousand adults; B.C. has between ten thousand and fifteen thousand, and Washington and Oregon together have about three thousand.

Populations fluctuate with the seasons. In B.C., for example, as many as thirty thousand adults and juveniles may inhabit the province in winter, when salmon are running. Groups of two hundred, five hundred, even a thousand may occupy one small area when food is plentiful. The eagle's habit of congregating on major spawning streams has earned it a reputation as a scavenger.

Although the eagle does feed on carrion, it is nonetheless a notorious hunter. From the top of a tree it selects its prey, perhaps a scoter, scaup, or grebe, from a flock floating on a bay or lagoon. One ill-fated straggler is singled out and the great white-headed predator lunges, forcing its prey to dive. The bird surfaces and the eagle strikes again, scarcely allowing time for a breath of air. The attack continues until the quarry is too exhausted to seek refuge in the depths of the bay. The eagle unceremoniously snatches its enfeebled victim from the water and leaves as quickly, as quietly, as it came.

It is not uncommon to see an eagle swoop down from a shoreside branch and clench an unwary fish swimming near the surface. If herring or other small fish are available, an eagle can devour one in midflight before zooming in for another catch. Larger fish, up to three kilograms, that swim near the surface are also fair game. The eagle circles overhead until a fish is sighted, then swiftly descends and sinks its talons into the flesh. Sometimes the fish is too heavy and the eagle, unable to return to flight, uses its wings to swim with the fish in tow. Occasionally eagles drown trying to make it to shore.

The bald eagle is associated almost exclusively with large bodies of water, and though many inhabit inland lakes and rivers, it is primarily a maritime bird. Eagles are seafood gourmets, proficient fishers and hunters that stalk rocky shorelines and beaches, probing seaweed and poking in tide pools for snails, shellfish, and crabs. Biologists who climbed into an eagle nest in the Queen Charlotte Islands were surprised to find 357 abalone shells.

While the eagle survives mostly as a hunter, fisher, and scavenger, it is also a well-known pirate. In many cases the loser is a hard-working osprey, an equally masterful fisher but a reluctant defender of its catch against a determined eagle. The eagle chases the osprey in flight, forcing it to drop its dinner for the eagle to retrieve. Sometimes eagles work in pairs: while one harasses the osprey, the other grabs the discarded fish as it falls through the air.

The bald eagle's unrivaled hunting skills are a paradox, the cause of both its salvation and its demise. Possibly the most versatile predator in North America, and undoubtedly the most resourceful bird of prey, the eagle was among the most persecuted raptors in North America until the early 1960s, when new laws declared it a protected species. Although the bald eagle is the emblem of the United States, Americans and Canadians alike considered this heraldic bird a pest: government-sponsored bounties were offered as a means of exterminating it. It is believed today that the number of bald eagles that survive outside B.C. and Alaska is smaller than the total number killed in Alaska in the days when the carnage was encouraged. Now their numbers are stable, possibly increasing, although bald eagles remain on the endangered list in the United States. Yet even today, the majority of injured eagles taken to wildlife infirmaries are suffering from gunshot wounds.

Eagles face other threats from man. Predators at the top of the food chain are vulnerable to environmental pollutants. Chemicals like DDT are found in seabirds, a source of food for raptorial birds. Aerial sprays to control forest insects kill raptors and their prey. One of the most menacing threats is the loss of nesting sites through logging: birds of prey have extreme difficulty reestablishing nests after disturbances.

Nests along the shores are an encouraging sight to eagle watchers. Partners for life, each couple returns — sometimes as early as February — to the same nest year after year, to lay one, two, or three eggs and share incubation duties. Only mature four- or five-year-olds, with distinctive white heads and tails, pair up to breed. An adult bird may measure almost a meter from its tail to its head and weigh between 2.5 and 6.3 kilograms.

Big birds build big nests — two or three meters across by fifty or sixty centimeters deep — usually within a hundred meters of the ocean, with commanding views over their hunting grounds. The nests, constructed within the crown of mature conifers, are platforms of beach sticks, dead branches, and twigs. The interiors are lined with soft materials such as seaweed, grasses, moss, and beach debris like plastic tarp fragments and frayed nylon line. Nests weighing a tonne are common, and many eventually topple from their lofty perches. Occasionally nests are made on cliffs; a few have been discovered on the ground on rocky offshore islands.

It was once thought that bald eagles traveled only short distances, but studies now have shown they are migratory. Migrations begin after the

BELOW: AN OSPREY FEEDING ITS CHICKS.

young have fledged, by September or October. The distances some eagles travel are astounding when one considers they were believed to be nonmigratory. Two eagles banded in the Chilkat River area of Alaska during late fall were found in late winter on Vancouver Island, some thirteen hundred kilometers away. In another survey, eagles wearing radio transmitters and banded in Washington's Skagit Valley flew about nine hundred kilometers to the Queen Charlotte Islands.

While the bald eagle is resourceful, the osprey is the epitome of form and precision. Known as the "fish hawk," the osprey survives almost exclusively on fish, spending all of its life within easy reach of large water bodies. It soars above the sea, searching for fish near the surface. When it sights one, it stops, and with tail feathers spread, hovers momentarily over its prey. Then, from a height of fifteen or thirty meters, it plunges feetfirst into the water, often disappearing completely. Other times it zeroes in on a fish and snatches it from the surface in one quick skim. Either way its deftness is proven by its success — it makes a kill more than half the time.

The osprey is the only bird in the world that has combined the use of its specialized raptorial feet with diving. Its large feet and talons are covered near the ends with sharp spicules, which help grip slippery fish. As a further aid, the cere, or horny lump at the base of the bill, is large enough to be drawn over the nostrils as the osprey dives.

Slightly smaller than eagles, ospreys are large hawks. Like eagles, they mate for life and lay their eggs in large conglomerations of sticks and grasses, which

ABOVE: ONLY MATURE FOUR- OR FIVE-YEAR-OLD BALD EAGLES DEVELOP THE WHITE HEADS AND TAILS.

81

are used repeatedly and repaired each nesting season. Their nests are generally more conspicuous than eagle eyries, standing out in the open atop broken snag trees, on utility poles, pilings, navigational aids, even construction cranes and chimneys.

About the same size as the osprey, the red-tailed hawk is more commonly seen near forests and open grasslands, perched on poles or fence posts, scanning fields for rabbits and rodents. Adults are easily recognized by their rufous tails: the tips of the feathers are light-colored with a dark band. Another giveaway is the harsh, prolonged scream, a familiar sound to seasoned birders.

Red-tails are opportunistic feeders: besides rodents and rabbits, they catch birds, snakes, and insects. Sometimes they watch from a perch, then swoop down on their prey. Other times they hunt on the wing, gliding low and snatching their victims on the move.

Far above most other birds of prey, the turkey vulture drifts effortlessly on afternoon thermal currents. As high as a kilometer in the air, it surveys a territory of three square kilometers or more, reaching speeds of sixty kilometers an hour with near-imperceptible movement of its wings. It stays aloft all day, descending only when it spots a dead deer, seal, sea lion, skunk, rabbit, or other carcass. When it sees a meal, it abruptly halts its listless soaring and glides straight for it, a signal to other vultures that dinner is served.

If the carcass is a large animal, the vulture approaches with trepidation, circling cautiously, sneaking in for a quick peck then recoiling as if escaping the strike of a venomous snake. Once satisfied the animal is dead, the bird gorges itself on the carcass. Some people consider the vulture's eating habits repugnant,

OPPOSITE: THE RED-TAILED HAWK OFTEN HUNTS IN OPEN FIELDS.

BELOW: OSPREYS, KNOWN AS "FISH HAWKS," SUBSIST ALMOST EXCLUSIVELY ON FISH.

but they serve a useful purpose by ridding the environment of potential health hazards.

The turkey vulture is not an especially engaging bird, with its featherless, purple-pink head, topped with a crop of prickly black stubble. Its feet aren't as strong as those of most raptors because, as a carrion feeder, it relies more on its sharp, curved beak to dismember its meal. On the wing, however, the turkey vulture floats with a graceful, mesmerizing rhythm, an enviable freedom we can only relish vicariously.

On the northwest coast these birds are partial to the open pastures, golf courses, and estuaries of the inside waters. They are abundant in Georgia Strait and on the Gulf and San Juan islands. During fall migration they gather in huge flocks, numbering in the hundreds, on the southern tip of Vancouver Island and offshore islands, where they roost in tall conifers.

In early spring a few vultures select nesting sites around southern Georgia Strait and eastern Juan de Fuca Strait. Nest locations vary and can be in hollow logs, stumps, or unused hawks' nests, often in forests of Douglas fir, arbutus, and Garry oak. Vultures prefer caves or ledges on cliffs or precipitous slopes, where they loosely assemble twigs, sticks, and other debris, and raise one or two offspring.

Ravens, Crows, Steller's Jays, and Kingfishers

The history of the raven is in some ways similar to that of the wolf. While old-world superstitions associated the big black bird with evil spirits and witchcraft, Indians held it in high esteem. It brought law and order to their societies, and created light by throwing pieces of shining mica into the air. The raven's prominence in legends and religion is depicted in the artistry of coastal natives. Good or bad, the mystical raven has always been evocative to man.

The hollow cluck of a raven as it soars overhead is like no other sound in the wild. It is one of many raven calls: early anthropologists believed Indians, who knew this bird as "Trickster," could forecast weather and predict events by interpreting the raven's vocalizations. Today researchers believe these sounds are used to communicate, and each has its own meaning.

The raven is a year-round resident of the Pacific Northwest, where it is found on seashores, in coniferous forests, and on mountain peaks. Like other members of the family Corvidae, it is a scavenger, feeding on carrion and garbage. It is also a predatory bird that wanders on foot, catching frogs, insects, eggs, and nestlings. Seasonal berries and fruit are an important part of the raven's diet.

The reasoning mind of the raven is legendary. It can remove lids from garbage cans, or trick other animals into forfeiting their food. It is common for one raven to goad a dog into chasing it, while another steals the dog's bone. At the seashore the raven opens oysters, clams, or mussels by dropping them from the air onto rocks or roads, a technique also used by crows and gulls.

In spring ravens pair up and build large nests in trees or on cliffs. The female incubates four to six eggs; both parents care for the nestlings. Where food is abundant, ravens may forage in groups, but they are mainly a solitary species and don't form large flocks like crows.

Except during summer breeding, northwestern crows fly in huge morning and evening flocks, as they commute between daily feeding grounds and offshore islands, predatorless havens where they roost safely for the night. The northwestern crow is indigenous to southern Alaska, British Columbia, and northern Washington. The slightly larger American crow is found in southern Washington and Oregon and elsewhere on the continent.

Like ravens, crows are intelligent and have an affinity for shiny objects. Coins, jewelry, fishing lures, or tin foil left unattended at a campsite are fair booty to a crow. Who knows what they do with it? Their amusing habits are often overlooked because, like ravens, crows are both scavengers and predators: they annoy devoted birders by causing serious harm to other, more colorful species. In seabird colonies these plunderous birds march boldly among the nests, grabbing unguarded eggs and hatchlings. In marshes, red-winged blackbirds protect their offspring by mobbing crows, distracting them from the nests.

On land, crows unintentionally help gardeners by eating insects, slugs, mice, and moles. Farmers who plant corn, however, are no friends of the crow. Marauding bands of North American crows descend on freshly planted corn fields and devastate the crops before they get a chance to germinate. Campers, often awakened at sunrise by loud, irritating caws, also show little affection for crows.

The raucous cackle of the Steller's jay, a smaller relative of the crow, is more pleasing to the ear. It is a signal that one of the more fascinating birds of the coniferous forest is nearby, and a warning to campers to hide their food. These robin-sized, iridescent blue freebooters shamelessly pilfer unguarded food at picnic tables and campsites. They are as fearless as they are curious, and the moment a camper's back is turned they zip down and make the heist —

LEFT: COURAGEOUS RAVEN STEALS A MEAL FROM A GRIZZLY BEAR.

ABOVE: THE NORTHWESTERN
CROW IS INDIGENOUS TO
SOUTHERN ALASKA, B.C., AND
NORTHERN WASHINGTON.

without even the courtesy of waiting until the campsite is deserted before beginning their thievery.

The only crested jay in the west, the Steller's was discovered in the mid-1700s on Alaska's Kayak Island by Georg Wilhelm Steller. A naturalist with the Vitus Bering expedition, Steller is credited with providing the world's first descriptions of fur seals, sea otters, and other North Pacific species. In 1987 the people of B.C. adopted the Steller's jay as their provincial bird.

These cocky jays are found from Alaska to Central America. Like crows, they feed on carrion, insects, eggs, nestlings, fruit, and berries. In the Pacific Northwest their favorite food is the Garry oak acorn. This oak species is found only on the west coast, from California to central Vancouver Island. Steller's jays carry acorns in their throat pouches and store them in caches for later use. It is thought that these birds are a major distributor of Garry oak trees. In residential areas, Steller's are enticed to feeders by peanuts and other large seeds, which they also hide. Recent research indicates that Steller's jays remember the locations of literally hundreds of food caches.

Steller's jays are not true migrators, but they move deep into the forests to nest. During nesting their personalities undergo a complete turnaround: their characteristic audacity changes to acquiescence, and they become secretive. These abrupt mood changes give the impression that they have migrated, when really they have just faded into the flora.

Another coastal chatterbox is the belted kingfisher. Its high-pitched clatter pierces the air as it flits from branch to branch in search of fish, tadpoles, frogs, and aquatic insects. It takes its prey by diving headlong from branches hanging over the water. When branches aren't handy, it momentarily hovers, like an osprey, before folding its wings and crashing through the surface. It grasps its prey in a long, heavy beak.

The kingfisher lives anywhere there is open water, from Alaska to the Caribbean. A nonmigratory bird, on the coast it maintains a territory in which most of its activities are carried out. Its nesting habits are as fascinating as its hunting routine. It burrows more than two meters into road cuts, stream or seashore banks and lays six or eight eggs at the end of the tunnel. No nest is built, but by the time the eggs hatch, an assortment of regurgitated fish bones makes an acceptable substitute.

BELOW: THE HARSH CALL OF THE STELLER'S JAY IS FREQUENTLY HEARD IN PACIFIC NORTHWEST CONIFEROUS FORESTS.

Shorebirds and Waders —
Shallow-Water Sleuths

Coastal wetlands are important not only to waterfowl, but to a multitude of shorebirds. They are vital resting and staging areas for hundreds of thousands of birds, many migrating as far as fifteen thousand kilometers through three continents between nesting and wintering grounds. In British Columbia, enormous flocks gather on mud flats and beaches on the northern Queen Charlottes, on Vancouver Island at Tofino and Comox, and in the Fraser delta. Of comparable importance are Grays Harbor and Willapa Bay in Washington; the lower Columbia River, Tillamook and Coos bays in Oregon; and Humboldt Bay in California.

From nesting sites in eastern Siberia and the northern tundra to winter wetlands in South America, these habitats are of international significance. The well-being of one affects the other, and each country is responsible for the maintenance of its own environments. In the Pacific Northwest, there is concern over Central American use of DDT and other toxic chemicals that are restricted in Canada and the U.S. Here at home, our track record regarding habitat losses, oil spills, and pollution is far from perfect. Through international agreements and wildlife management plans, shore and wetland rehabilitation, and outright purchases of critical habitats, we are correcting errors of the past and finding new ways to avoid future mistakes.

About fifty species of wading birds and shorebirds — herons, oystercatchers, plovers, sandpipers, phalaropes, and others — occupy our beaches and salt marshes at various times of year. Just as the bald eagle is the most conspicuous raptor, the northwest's most noticeable wading bird is the great blue heron, and for the same reasons — numbers and size. Of the six local heron species, the great blue is the most abundant and widely distributed.

Standing as tall as a six-year-old child, they are easy to recognize, with their spindly, knobby-kneed legs and long necks hunched into their shoulders like grumpy old men. Coastal Indians personified the heron as old and wise. Herons are exceptionally patient, staring endlessly into shallow lagoons, awaiting the right second to strike an unsuspecting stickleback or flatfish with long, spearlike bills. They seem most successful in eelgrass and tidal marshes on the seashore; in the uplands they hunt in streams, backyard ponds, sloughs, and irrigation ditches for fish, frogs, and mice.

Indians also portrayed the heron as a loner, yet groups of half a dozen or more are common. In the sandy-bottomed shallows of the Fraser delta, as many as three hundred have been counted probing the eelgrass. These big, steel-gray birds are found across the continent and down to South America.

OPPOSITE: THE BELTED KINGFISHER SELECTS ITS PREY FROM PERCHES OVER THE WATER.

BELOW: SEMIPALMATED SANDPIPERS ARE AMONG THE SMALLEST SHOREBIRDS.

ABOVE: A GREAT BLUE HERON
SNATCHES A FISH IN ITS
SPEARLIKE BEAK.

Herons in the northwest don't migrate long distances but disperse along the coast from breeding colonies in late summer. Heronries of one or two to two hundred nests are located in groves of alder or other trees close to feeding grounds. Nests, near the treetops, are sizeable collections of twigs and branches, and the ground below is liberally spattered with heron guano. They arrive in March or April and stay through summer. Newly hatched offspring are grouchy-looking little beasts, with fuzzy rooster tails, long scraggly down, and glaring, contemptuous eyes.

Another bird that stays close to home year-round is the black oystercatcher. As large and as dark as a crow, the oystercatcher's long, crimson bill matches the rings around its yellow eyes.

Found only near the sea, the oystercatcher is most often seen on rocky shores, islands, and islets. It also feeds in lagoons, and on gravel shores and mud flats, but avoids beaches when the surf's up.

Its well-camouflaged nests of stones and shell chips are built almost exclusively on offshore islands, amid drift logs, or on rocky, grass-covered promontories. Nests are often surrounded by white shell fragments, perhaps to reflect the sun and prevent overheating of developing eggs. Nests too close to the sea are frequently washed away by storms or extreme tides. The oyster-catcher is widely distributed on the coast: breeding populations are estimated at seventeen hundred pairs in Alaska, a thousand in B.C. and two hundred in Washington. Like the heron, this species disperses from nesting grounds in late summer. It gathers in flocks of a hundred or more in early fall.

The most abundant shorebird on rocky coasts is the black turnstone. Flocks of four thousand have been recorded on the north coast near the start of the autumn migration in July, and as many as two thousand at a time appear on the east coast of Vancouver Island in winter. Though its range is broader than the oystercatcher's — from Alaska to Mexico — it shares similar habitats. It scampers along gravel and rock shores, overturning pebbles with its bill to find crabs and marine crustaceans. It also feeds on reefs and floating kelp beds. The turnstone, which breeds mainly in southern and western Alaska, doesn't build a nest: it simply scrapes a depression in the ground and lays about four eggs.

A part-time shorebird in the northwest is the semipalmated plover. Smaller than a robin, larger than a sparrow, it is one of four plover species to appear in the northwest. It is distinguished by its single black breast band and eye stripe, and short yellow bill with a black tip. The name semipalmated, which means "possessing half a palm," is descriptive of the small web which joins its three front toes. It hunts tiny marine animals in the intertidal zone by dashing along the shore, stopping abruptly, and quickly jabbing the sand for a beach hopper or crab.

After wintering as far down the coast as South America, "semipalms"

Healthy Heron, Healthy Home

Great blue herons, because of their high numbers and wide distribution throughout the northwest, are good indicators of the health of our shores, estuaries, and wetlands. Studies of heron eggs in Georgia Strait during the 1980s show that while levels of chlorinated hydrocarbons are falling, dioxins and furans are rising, at least in some areas. Dioxins damage heron embryos and the failure of a colony to fledge young birds in 1987 was tied to dioxin levels: the nests were adjacent to a pulp mill.

Although nesting herons are sensitive to disturbances, they nonetheless breed in surprisingly busy places — downtown Vancouver's Stanley Park, for example. They are known to abandon colonies, however, when upset by logging, construction, or off-road vehicles. Herons are easily spooked by people walking beneath their nests: they utter deep, guttural croaks and nervously hop between trees, keeping a watchful eye on intruders.

The Misnamed Oystercatcher

The black oystercatcher, with its long, crimson bill, is well equipped to eat shellfish. The end of its bill is vertically flat, a tool to insert into oysters, mussels, and other bivalves and sever the adductor muscle, forcing the shells apart. Although it can gobble half a dozen oysters in an hour, the name "oystercatcher" is somewhat of a misnomer because it feeds more on mussels, limpets, crustaceans, and sea worms.

This pretty bird is a bit of an oddball, scooting along rocky shores, reefs, and islets, uttering loud, high-pitched peeps as it pokes in pools and under seaweed in the intertidal zone.

stop here in spring en route to northern nesting areas. Aggregations of ten to fifty birds are common, and flocks of two thousand may form at staging sites such as Long Beach in Pacific Rim National Park. A few nonbreeders stay behind for summer, but most fly to Alaska and Canada's arctic tundra. Nests are also found along the B.C. coast. In the northern Queen Charlotte Islands, semi-palmated plovers scratch out small depressions in the sand and line them with seaweed, grass, clam shells and pebbles. They reappear on the south coast in late summer and early fall while returning to warmer equatorial climes.

The slightly larger black-bellied plover is far more numerous here during fall migrations. Flocks of seven thousand have been recorded at Boundary Bay, B.C., and the Fraser delta accommodates Canada's largest wintering populations. It associates with similar-sized shorebirds, such as turnstones and golden plovers, on tidal mud flats, sandy beaches, islets, farms, and golf courses.

Like the lesser golden plover, the black-belly is a long-distance migrant, wintering from Alaska to South America, breeding in the Canadian arctic and Siberia. The golden plover is much rarer in the Pacific Northwest at all times of year.

The dunlin, a few feathers larger than a semipalmated plover, is one of the northern hemisphere's most abundant shorebirds. It is widely distributed along the coast, from Alaska to Mexico. Flocks of fifty thousand or more converge on the extensive mud flats of Boundary Bay in fall and winter. At times in November more than a hundred thousand dunlins have been counted in the Fraser delta.

The seasonal movements of dunlins distinguish them from other sandpipers. According to recent research, those that breed in northern Alaska winter on Asia's Pacific coast; western Alaskan breeders spend winter along the west coast of North America. Its southward migration also happens later: a few arrive in the Pacific Northwest in the height of summer, but the migration peaks in late October and continues into December.

Another unusual trait of dunlins is their method of defending themselves against hawks. When feeding at low tide they spread along the shore: when a hawk flushes them, they take to the air and gather in a tight group, twisting and turning as they fly. The large, flying mass they create discourages a hawk from attempting an assault.

The migration habits of the short-billed dowitcher are quite the opposite of the dunlin's — it is one of the earliest southbound migrants, appearing on northwest shores in July and early August, heading for wintering grounds from the southern U.S. to Peru and Brazil. By October all but a few stragglers have gone and, with the odd exception, none remains for winter. They are most

LEFT: SEMIPALMATED PLOVER, ONE OF FOUR PLOVER SPECIES IN THE NORTHWEST.

BELOW: THE BLACK OYSTERCATCHER'S LONG BILL IS SPECIALLY DESIGNED TO OPEN BIVALVES.

ABOVE: BLACK-BELLIED PLOVERS
FORAGE ON BEACHES AND MUD
FLATS.

LEFT: BLACK TURNSTONES ARE
PLENTIFUL ALONG ROCKY
SHORES.

97

noticeable during spring migration, when they stop en route to breeding sites in Alaska and northwestern B.C. Flocks in the thousands gather on the shores: fifteen thousand have been counted on the mud flats of Tofino Inlet.

The southward migration of the larger long-billed dowitcher, which winters from the southern U.S. to Guatemala, coincides more with that of the dunlin. The exodus from its Siberian and Alaskan nesting grounds doesn't peak on the northwest coast until September or October and dwindles into December. While the long-bill frequents tidal mud flats and sandy shores, it is also partial to swamps, flooded fields, and ponds.

ABOVE: LONG-BILLED DOWITCHER PROBES MUD FLATS WITH ITS SHARP BEAK.

OPPOSITE: SOUTHBOUND MIGRATING SHORT-BILLED DOWITCHERS APPEAR IN THE NORTHWEST AS EARLY AS JULY.

Western Sandpipers

The sparrow-sized western sandpiper is the most abundant shorebird in the Pacific Northwest. Flocks as large as a hundred thousand pass through during fall migration, which begins as early as the third week of June. This southbound migration peaks in mid-July and continues through October, with a few stragglers in early November. The world's entire population of 1.5 million stops at both the Fraser delta and Grays Harbor on its southbound flight. The western sandpiper is scarce in the northwest through winter, when it prefers the balmy breezes of southern California or Peru.

One of the smallest sandpipers, these little birds seem always at the edge of the sea, moving up and down with the tides. They scuttle along below the tideline, poking in the sand, fleeing en masse at the slightest intrusion. At high tide they retreat up the beach and preen themselves until the ebbing tide reveals more food.

Western sandpipers, along with semipalmated and least sandpipers, are informally known as "peeps" because of their frequent, high-pitched, single-syllable squeaks.

Both dowitcher species have long bills which they use to probe for worms and crustaceans. They insert their beaks deep into the mud and, like a sewing machine, unearth their prey in a series of rapid up-and-down thrusts.

Feeding shorebirds prefer walking to swimming, but the greater yellowlegs often wades beyond the depth of its skinny legs and paddles back to the shallows. This bird travels in small flocks through the northwest, where it's an early-

spring migrant, with the first groups moving through to Alaska and northern B.C. in late February or March. It nests at elevations of about a thousand meters, in wet, sparsely treed forests near sloughs or bogs. In autumn the southbound migration is just about over by early November. Small groups, usually fewer than one or two dozen, remain at southern Vancouver Island, Washington, and Oregon through winter.

Seabirds – Gulls, Cormorants and Alcids

SEABIRDS — MILLIONS UPON MILLIONS — congregate in great flocks and feed in the plankton-rich waters over the continental shelf. At least 6 million breed throughout the northwest; another 40 million nest in Alaska. The numbers at individual nesting colonies are equally astounding: 1 million at Triangle Island, off northwest Vancouver Island; forty-four thousand on Puget Sound's Protection Island;

twenty-four thousand at Destruction Island off the Washington coast; eighty thousand each at Oregon's Goat Island and Three Arches Rock.

About 80 percent of the nesting seabirds in the Pacific Northwest are alcids — murres, murrelets, auklets, puffins, and pigeon guillemots. Most nest in burrows or on bare rockfaces on the cliffs of offshore islands, but some, such as the marbled murrelet, build nests in the tops of mature conifers. Other burrow nesters are the storm-petrels, which comprise about 15 percent of our breeding seabirds. All of these species are generally found at rugged, isolated colonies: huge populations nest on the Queen Charlottes, and on the Scott Islands, off northwest Vancouver Island. Cormorants, which make up about 1 percent of

the breeders, build nests of sticks and seaweed in trees, or on guano-stained cliffs. They occupy colonies on the outer coast and inside waters, often close to civilization.

The omnipresent gull, surprisingly, makes up only 2 percent of the breeding seabirds. Although more than a dozen species scavenge Pacific Northwest shores, only one — the glaucous-winged gull — nests on the coast. The gull's ubiquity can be attributed to its penchant for human garbage: gull populations, like our mounds of household debris, are growing in heavily populated areas such as southern Georgia Strait. Like other seabirds, however, gulls prefer fresh seafood to garbage.

ABOVE: A HERRING GULL TAKES FLIGHT.

Our phenomenal populations of seabirds are closely tied to concentrations of fish and plankton, which are produced in association with currents. In the big picture, the North Pacific Current swirls in a counterclockwise direction toward the northwest coast, where it splits. One current then moves north into the Gulf of Alaska, the other travels south, becoming the California Current. These currents influence overall oceanic conditions in the North Pacific.

Closer to shore, food appears in turbulent channels, rip tides, and back eddies along the continental shelf. Near the edge of the shelf, at depths of two hundred to three hundred meters, areas where upwelling of colder, deeper water occurs are particularly productive. These are places where a combination of the earth's rotation and winds force cold, nutrient-laden water to the surface from depths of one hundred to three hundred meters. In the sunlit waters near the surface, phytoplankton — the primeval food of the sea — rapidly reproduce by photosynthesis. Phytoplankton, or minute plants, are food for zooplankton, or minuscule animals, which are prey for fish and seabirds. So, logically, plankton, fish and seabirds congregate in the same places.

Out on the continental shelf, shallow banks which rise to within forty or fifty meters of the surface are especially fishy. Off southwest Vancouver Island, noisy flocks of seabirds gather at Swiftsure, Finger, Gullies, La Perouse, 6 Mile, and Amphitrite banks. Farther north, in Queen Charlotte Sound, large numbers of birds feed at Cook, Middle, Goose Island banks and others.

Good feeding areas are also located near deep underwater canyons that intrude into the edge of the continental shelf, much like the fjords that penetrate the mountains of western Vancouver Island and the mainland. The Juan de Fuca Eddy, just off the mouth of Juan de Fuca Strait, is a huge gyre that draws nutrients up from a deep canyon and spins them into the middle of the eddy.

These fertile feeding waters support sizeable fishing fleets. While some birds, mainly gulls, feast on offal tossed overboard by fishermen, diving birds risk entanglement in nets. One fisherman reported that five hundred ancient murrelets drowned in his gill net off the west coast of the Queen Charlottes.

Other fishing farther offshore affects birds that nest in the Pacific Northwest and elsewhere in the North Pacific. Each day for half the year Asian fishermen set out enough monofilament drift net in the mid-Pacific to encircle the earth. They are fishing for squid, but their indiscriminate methods kill 750,000 seabirds a year. Similar curtains of death, used by the Japanese to catch salmon, drown about 200,000 seabirds a year in the international waters of the North Pacific. Canada and the U.S. have banned drift nets in their 370-kilometer offshore economic zones. But in the lawless no-man's-land beyond reach of North American statutes, only goodwill among nations can stop the slaughter.

Another man-made, and rather appalling, danger on the high seas is floating plastic. It is estimated that every day the open Pacific is littered by between one thousand and four thousand pieces of plastic per square kilometer, much of which is carried to nesting colonies or washed onto Pacific Northwest beaches. Small particles, which are mistaken for food, are hazardous to any seabird: tube-nosed swimmers — shearwaters, petrels, fulmars, albatrosses — are most susceptible. With smaller gizzards than other species, they are unable to regurgitate swallowed bits of plastic. The plastics impair digestion, block gastrointestinal tracts, cause starvation and hormonal harm, and increase the intake of PCBs. Nesting birds able to bring up ingested plastic inadvertently feed it to their offspring.

BELOW: DOUBLE-CRESTED CORMORANT GATHERS NEST MATERIALS.

The greatest threat to seabirds is oil spills. Since the development of Alaska's oil fields in the late 1970s, people of the northwest coast have dreaded the day the inevitable would happen. And not one, but two catastrophic spills occurred within three months of each other. On Dec. 22, 1988, a tug collided with its barge, the *Nestucca*, off Grays Harbor, Washington, dumping some 875,000 liters of Bunker C oil — a syrupy, black tar — into the sea. Nearly thirteen thousand dead seabirds, mostly murres, were collected from the beaches of Washington and Vancouver Island. No one knows how many others died.

Then on March 24, 1989, the supertanker *Exxon Valdez* grounded on a reef in Alaska's Prince William Sound, disgorging 42 million liters of oil through its punctured hull. According to court documents filed after the spill, nearly 600,000 seabirds, mostly murres, were killed.

The majority of the continent's oil and gas production occurs near the North Pacific coast: tankers like the *Exxon Valdez* will ply northwest shipping routes for many years to come. In fact, traffic could increase if a moratorium on petroleum exploration off the coast of B.C. is lifted. Since the late 1960s, oil companies have drilled more than a dozen exploratory wells in Hecate Strait and Queen Charlotte Sound, and off the west coast of Vancouver Island. Small deposits of gas and oil were found, but in 1972 the Canadian government ordered the moratorium.

Twelve years later, with the drilling ban still in place, a panel was appointed to assess the environmental, economic, and social effects of renewed petroleum exploration. If drilling brought favorable results, oil and natural gas production could begin ten or fifteen years later. In 1987, the B.C. and Canadian governments were negotiating terms for lifting the moratorium. But after the *Nestucca* and *Exxon Valdez* spills, it was announced the moratorium would remain in effect at least until 1994.

Those two spills were specific, and easily documented. What is unknown is the extent of oil pollution caused by the illegal pumping of bilges at sea, by discharges from urban storm drains, by spills from pleasure boats. Researchers believe so-called minor spills kill maybe half a million birds a year off the west coast of North America. Even small spills are deadly: like fish, oil moves with the currents, accumulating in the same places that birds feed. Diving birds go

BELOW: MEW GULL, ONE OF THE SMALLER GULLS ON THE COAST.

down through the oil; they come up through the oil.

To monitor the impact of minor oil spills, volunteers in British Columbia have been gathering oil-killed seabirds from beaches since 1990. Similar surveys in New Zealand, the Netherlands, and the Shetland Islands have prompted greater surveillance of ships, as well as changes in regulations concerning the loading and movement of oil carriers.

Natural phenomena also take their toll. Violent Pacific storms have blown storm-petrels more than five hundred kilometers inland. In 1970 an estimated sixty-eight thousand common murres died at Bristol Bay, Alaska, due to starvation precipitated by inclement weather. Hatchlings succumb to hypothermia under pitiless coastal rains. Cloudy weather can reduce summer phytoplankton blooms during nesting seasons, when they are needed most.

Predation is another killer. A single peregrine-falcon or bald-eagle family might take a thousand seabirds a year. River otters accost storm-petrels as they return to their burrows at night. Seabirds nest on offshore islands and islets to avoid predation. However, introduced predators — rats, racoons, ferrets, foxes, feral cats, or dogs — can be devastating. On Langara Island, the most northerly

of the Queen Charlottes, the breeding population of ancient murrelets was about 180,000 in 1971. A decade later it had dropped to 45,000: biocides or a reduction of food caused by unusually warm water could have contributed to the decline, but biologists suspect introduced black rats were largely responsible. Racoons, recently released on nesting islands in the Charlottes, have begun to swim to new islands to feed on eggs and nestlings. Roaming cats and dogs belonging to lighthouse keepers have wiped out entire colonies of nesting birds. Even researchers attempting to help seabirds accidentally squash nests, or scare away doting parents when they approach nestlings. Helicopter landings frighten breeding birds, causing many to abandon their nests.

Despite these numerous perils, the seabirds of the Pacific Northwest appear to be in good health. Populations fluctuate, but, at least for now, there's reason for guarded optimism about the future of our seabirds. As with other marine animals, humans hold most of the cards.

ABOVE: THE CALIFORNIA GULL IS THE MOST ABUNDANT GULL ON THE COAST IN AUTUMN.

Gulls and Terns —
The Nebulous Sea Gull

The term "sea gull" is nebulous. To the casual beachcomber it could mean any of a dozen or so species. A flock of "sea gulls" might be an assortment of California gulls, herring gulls, glaucous-winged, western, or Thayer's gulls. They look alike, they sound alike, they hang out by the sea. They must be "sea gulls."

Not so. Taxonomically, there's no such thing as a simple "sea gull." If, however, this honorific were bestowed upon one species, it would be the glaucous-winged gull, the only bird of its kind to spend its whole life on or around the sea. This is the proverbial "ubiquitous sea gull." Huge flocks amass on bays and beaches, lakes and lagoons, farmers' fields, city parks, and garbage dumps. They wheel and squeal over schools of herring or sandlance, and boisterously beg for scraps from canneries and fishboats. Glaucous-winged gulls travel far upstream from the sea to stand beside eagles and peck holes in spent salmon. Flocks of twenty-five thousand are not unknown.

When it gets the urge to propagate, the glaucous-winged gull takes up residence with cormorants, murres, and other nesters on offshore islands and islets. At least half the nests in the Pacific Northwest are in colonies along the inside waters, throughout Georgia Strait and Puget Sound. Nearly 40,000 breed in Washington, 60,000 in B.C., and 270,000 in Alaska. With growth in human populations — and our never-ending accumulations of garbage — glaucous-winged gull numbers around Juan de Fuca and Georgia straits have more than tripled in the last fifty years. They have begun to expand beyond traditional nesting sites to such unlikely spots as city rooftops, window sills, busy shipyards, or support beams of highway bridges. Perhaps five hundred or more now nest in downtown Vancouver.

Only the glaucous-winged gull breeds on the seashore: the others move to lakes, streams, and ponds. Among the freshwater nesters is the California gull, which outnumbers all the other gulls on the ocean in autumn. Staging areas on the outer coasts see flocks of five thousand, or even ten thousand. A few stay for winter, but most move to sunny southern shores. They reappear

Avian Piscatologists

A study from 1979 to '81 on the effects of fish-eating birds on salmon fry at Vancouver Island's Qualicum River showed substantial differences between the feeding habits of the Bonaparte's gull and the glaucous-winged gull, both numerous and major predators. Feeding on fry released from a hatchery, the Bonaparte's worked only on certain tides in tight-knit groups where fish populations were densest. The glaucous-winged gulls were more loosely spread along the stream and there seemed to be no correlation between tides and feeding. Of a total of eight fish-eating bird species under study, Bonaparte's gulls were most efficient, taking as many as 290,000 fish fry one year, compared to the glaucous-winged gulls' catch of up to 4,000.

in spring as they head for interior mating grounds as far off as Saskatchewan or North Dakota. This bird became dear to the hearts of Mormon settlers near Utah's Great Salt Lake after huge flocks arrived and devoured a swarm of locusts that threatened the Mormons' first crop.

The smaller mew gull prefers cultivated fields to offshore fish schools in winter. Thousands follow farmers as they churn up earthworms and insects with their plows. These birds are nonetheless able fishers, taking herring and bait-fish from near-shore waters. Although overall numbers during autumn migrations may not be as high as California gulls, mew gulls gather in flocks of four thousand, six, or even seven thousand. In November, 1983, an estimated thirty thousand were seen feeding along a three-kilometer tideline near Victoria. Being smaller, the mew gull is less aggressive than other gulls and avoids territorial disputes. It also flies as far as Saskatchewan to breed, but unlike the California gull, it stays in the northwest through winter. While many migrate south, large numbers remain in Juan de Fuca and Georgia straits.

The tiny Bonaparte's gull, with its distinctive black hood, is an adept fisher. Its high-pitched screech carries across open water as it swarms over areas of upwelling and tide rips, repeatedly diving for fish. This graceful little bird is known as the "coho gull" because it feeds on small herring and sandlance forced to the surface by coho salmon. During the autumn coho season, Bonaparte's gulls number in the tens of thousands in northwest waters. They are followed by jaegers looking for easy meals. This gull is also an efficient predator of fish fry in salmon streams.

Large flocks remain here through winter, while some fly south to Mexico. They are most numerous in the northwest in spring as they pass through en route to inland mating grounds. Unlike other gulls, it nests in trees near ponds, lakes, alpine marshes, and muskeg.

Bonaparte's gulls are often confused with terns because of their black heads. The most famous of the Pacific Northwest's five tern species is the arctic tern, celebrated for its incredible annual migration between the world's two polar regions, a return journey of thirty-five thousand kilometers.

The arctic tern, however, is not as frequent a northwest visitor as its closest look-alike, the common tern. The best time to see these slender seabirds is autumn, when they rest on kelp beds or floating driftwood as they migrate from interior nesting sites to wintering grounds as far away as South America.

Terns are flamboyant fishers, hovering at heights of ten or twelve meters over the water, then plunging in for their prey. Like Bonaparte's gulls, they are plagued by jaegers: when they arrive in late summer and fall they are invariably accompanied by jaegers. Both are agile fliers and the aerial acrobatics of a jaeger in pursuit of a tern are spellbinding.

Cormorants — Greedy or Voracious?

The Funk and Wagnalls dictionary defines cormorants like this: "1. Any of various large, web-footed aquatic birds of wide distribution, having a hooked bill and a pouch under the beak in which it holds fish. 2. A greedy or voracious person."

The dictionary's second, unflattering definition isn't meant to slander this amiable seabird, but to attach certain avian connotations to rapacious people. While it is true the cormorant's appetite is insatiable, the bird's voracity isn't

ABOVE: GLAUCOUS-WINGED GULLS OFTEN SHARE NESTING SITES WITH CORMORANTS.

OPPOSITE: COMMON TERN PREENS ITS FEATHERS.

113

BRUCE OBEE

born from greed, but from a need for energy. Many seabird species, which feed irregularly, store energy for swimming and flying in body fat: the cormorant carries little fat and therefore must feed every day. Early mariners often ate low-fat cormorant meat, but the bird's ability to retain large amounts of oxygen in the blood gives the meat a redness, and rankness, inedible to even the coarsest palate. Hanging the bird from the rigging for a couple weeks, however, reduced the raw-fish fetor to an acceptable, if not delectable, standard.

The need for reliable food sources compels cormorants to live near cold, productive waters. Every morning long lines of cormorants, flying just above the surface, head to the fishing grounds for the day, then return to their roosts at dusk. With small wings and big, webbed feet set far back on their bodies, these dexterous divers are designed to chase even the fastest fish. Holding their wings tight against their bodies and propelling themselves with their oversized feet, cormorants commonly dive to thirty meters. Some have been caught in nets at fifty-five meters.

ABOVE: NESTING PELAGIC
CORMORANT.

For centuries Oriental fishermen kept flocks of cormorants to do their work. The birds, wearing rings around their long necks to prevent them swallowing the fish, were set free from boats and guided back by long lines affixed to the rings. When the fishermen had enough food for their families, the cormorants were permitted to feed themselves.

The powerful, streamlined body of the cormorant is a disadvantage when out of its aquatic element. It literally trips over its own big feet, and crushes its eggs and offspring. Its muscular hips and thighs make its body too heavy for its small wings. Through some evolutionary blunder, its dense plumage is not well waterproofed: a cormorant standing on a piling, beach or cliff, wings spread as though worshiping the sun, is airing them in the breeze, often in preparation for takeoff.

Getting airborne is a tricky task: it clumsily slaps its feet and wings along the water, as if trying desperately to push itself off from the surface. Once aloft, it laboriously beats its wings to keep from falling out of the sky. On land it takes off into the wind: astute observers notice that cormorant roosts and nesting colonies are always on unsheltered sides of islands and islets, facing prevailing winds.

Cormorants are sometimes called "sea crows" because of their dark black plumage. But a closer look at the three Pacific Northwest species reveals a variety of colors. The adult pelagic cormorant is actually a deep metallic green, with two small crests, green eyes, and a red throat patch. A young double-crested

cormorant is dark bronze, while an adult is black with two unobtrusive tufts on either side of the head, and a highly-visible yellow throat pouch. The iridescent green Brandt's cormorant has a cobalt-blue throat patch and eyes.

The cormorant's colorful throat patch is used not only to store fish but to attract a mate. A male seduces a partner by standing in the open — its head thrown back to expose its throat — and furiously flapping its wings.

There is no shortage of cormorants in the northwest, especially in Georgia and Juan de Fuca straits. Most plentiful and widely distributed is the pelagic cormorant, with over four thousand nesting pairs in B.C. and twenty-five hundred in Washington. It nests occasionally in sea caves, more commonly on steep, rocky offshore islands.

Mandarte Island lies on the B.C. side of Haro Strait, near the Canada-U.S. border, where Georgia and Juan de Fuca straits meet. It is typical of many nesting sites: from a distance it appears as a rocky hump rising from the sea, whitewashed with generations of seabird guano. A closer inspection shows hundreds of nests surrounded by the clamorous activity of thousands of breeding birds. Big nests of sticks, bones, and seaweed cling precariously to the cliffs, jealously guarded by their occupants. Each is located just beyond pecking distance of the next, and waddling only a few steps from home invites confrontation.

With 550 pairs of pelagic cormorants, it is one of the northwest's most important colonies. Pelagic cormorants, however, aren't the only birds to benefit from Mandarte's isolation. Nearly as many double-crested cormorants nest here, along with at least 1,000 pairs of glaucous-winged gulls, about 120 pairs of pigeon guillemots, and 150 pairs of song sparrows. It is also the only site in Georgia Strait where tufted puffins and rhinoceros auklets breed.

Only six kilometers from Vancouver Island's populated Saanich Peninsula, Mandarte is a convenient study site for scientists. Passing boaters see their shanties and viewing blinds, where devoted researchers spend hour upon hour spying on the birds, inhaling the fowl air. Studies of Mandarte's prolific birdlife began in the late 1950s, and it is said the island has produced more PhDs per hectare than any other piece of real estate in the northwest.

The birds of Mandarte have been of service to humans much longer than half a century. For centuries before scientists learned about the island, Saanich Peninsula Indians gathered gulls' eggs and camas bulbs from Mandarte every summer. By watching the moon and tides, they knew when the eggs would be fresh. Only dull eggs would be taken;

Cormorants and Herring

Like many seabirds, cormorants are especially fond of herring. The three here are resting on log booms in Vancouver Island's Northumberland Channel, where thousands of sea lions join the birds for the annual herring feast from late fall to early spring.

Although nesting Brandt's cormorants are comparatively scarce in the northwest, they are abundant here when the herring spawn. As many as four thousand Brandt's have been seen on Washington's San Juan Island in September, and seven thousand have been counted in the turbulent waters of B.C.'s Active Pass.

The total wintering population in inside waters is probably at least fifteen thousand birds. These big black fishers spread out in long lines and repeatedly dive for fish. Most of these wintering Brandt's cormorants nest in southern colonies, the largest at California's Farallon Islands.

BRUCE OBEE

older shiny ones were left to hatch, ensuring a continuous supply.

Pacific Northwest boaters are often surprised at the sight of double-crested cormorants nesting in dead trees. Like pelagic cormorants, the number of double-crested cormorants has increased dramatically in the northwest since the early 1900s. Today there are more than two thousand breeding pairs in B.C., about seventeen hundred in Washington.

Compared to double-crested and pelagic cormorants, nesting Brandt's cormorants are scarce in the northwest. B.C. may have fewer than 60 pairs, while Washington's total population may be 270 pairs.

Alcids — Most Plentiful Seabirds

Alcids are the true seafarers of the avian world. Many spend almost their entire lives beyond the horizon, drifting the storm-tossed waters of the North Pacific. In spring they gather in gigantic flocks for the annual pilgrimage to land. They move en masse to craggy, isolated isles for a hectic summer of parenting, carrying out many of their procreative duties under cover of darkness. When youngsters have fledged, colonies are abandoned; the birds return to sea with their offspring, which soon are left to fend for themselves.

With their chubby bodies and stubby necks, black and white plumage, stumpy wings, and upright stance, alcids are the northern hemisphere's answer to penguins. Though smaller, their methods of "flying" underwater are similar:

OPPOSITE: PARAKEET AUKLETS IN

ALASKA'S PRIBILOF ISLANDS.

118

steered by their webbed feet, short, powerful wing-strokes propel them to depths of thirty-five or forty meters.

The nine alcid species of the Pacific Northwest comprise 80 percent of our nesting seabirds, yet many are seldom seen. Their colonies, some populated by a million birds, are remote, usually steep, and often accessible only by helicopter. Common and thick-billed murres, pigeon guillemots, marbled murrelets, rhinoceros auklets, or horned puffins feed in shallow water relatively close to shore, but their foraging grounds may be some distance from human settlements. Others — Cassin's auklets, ancient murrelets, or tufted puffins — might spend the winter a hundred to two hundred kilometers offshore, feeding on plankton and fish.

Half of the northwest's seabirds are Cassin's auklets. The most widespread of all alcids, this robin-sized bird now occupies virtually all of its usable habitat. It is most plentiful in British Columbia, where the province's 1.2 million nesting pairs are about four-fifths of the world's total population. Some 800,000 breed on the Scott Islands; thousands of others nest on the Queen Charlottes. Neighboring Alaska has about 236,000 pairs, and 44,000 breed in Washington.

The Cassin's auklet is an offshore bird, feeding in the currents and upwellings at the edge of the continental shelf. Sightings in its ocean domain are infrequent but, during one April, more than eleven thousand were counted in less than three hours at Queen Charlotte Sound. It is slightly more visible at its nesting colonies, but it approaches its burrow only after nightfall. A single egg is laid in a burrow that may extend one or two meters into the ground.

Cassin's and rhinoceros auklets frequently share nesting colonies. The 340,000 pairs of rhinoceros auklets in B.C. comprise about 35 percent of the world population. Outside B.C., numbers are much lower, possibly 30,000 pairs in Washington, 60,000 in Alaska. Distinguished from the Cassin's by the prominent horn on its beak, the rhinoceros auklet is larger, about the size of a pigeon.

The parakeet auklet, with gleaming white eyes and an orange, parrotlike beak, breeds on the northern islands of Alaska and Russia. Another offshore bird, it hadn't been recorded in B.C. until 1988, when the *Nestucca* oil spill blackened the waters from Grays Harbor, Washington, to Vancouver Island. Fifteen parakeet auklets made history by coming ashore on Vancouver Island. Unfortunately, they were dead.

The common murre is more susceptible to oil contamination than other seabirds. Most of the dead birds recovered after Alaska's *Exxon Valdez* oil spill were murres. In the aftermath of the *Nestucca* spill, 80 percent of the oiled birds collected in Washington were murres; in B.C. they accounted for 42 percent. This bird, weighing more than a kilogram, is the largest alcid. It is also plentiful in the northwest: 15,000 pairs in Washington, 2,800 in B.C., 940,000 in Alaska. A hundred thousand may fish the inshore waters of Juan de Fuca and Georgia straits in winter.

Biologists have difficulty counting these birds because they don't build nests; a single egg is plopped on a barren shelf sticking out from a precipitous cliff. For the contents of the egg, the future is precarious. Gusty winds blow eggs into the sea; adults fleeing from predators knock eggs — or offspring — off the shelf. A small advantage is that murre eggs are pear-shaped, which minimizes chances of them rolling off cliffs.

The ancient murrelet, with graying, old-man plumage on its back, seems a bit of an oddball in the seabird fraternity. A nocturnal bird, it usually lays two eggs in burrows excavated in old-growth forests close to the ocean. When only two or three days old, a chick escapes the clutches of peregrine falcons, eagles, gulls, and other predators by making a quick exit to the sea at night, sometimes accompanied by its parents, ofttimes alone. By dawn the young hatchling, having never been fed by adults, may be thirty or forty kilometers offshore. Two years later it returns to the land, for the first time, to produce its own brood.

OPPOSITE: PIGEON GUILLEMOTS, YEAR-ROUND RESIDENTS, ARE LIKELY TO BE SEEN NEAR COMMUNITIES, AS THEY ARE SHORELINE FEEDERS THAT FREQUENT BAYS AND HARBORS.

Three-quarters of the world's ancient murrelets inhabit B.C. waters. There are 272,000 pairs in B.C., 85,000 in Alaska. Periodic sightings of ancient murrelets as far from the Pacific as the Great Lakes perplex researchers. The birds are always found there after severe storms, but no one knows how they could possibly be blown so far from home.

Some birders believe the marbled murrelet is the most fascinating alcid. With its furtive ways and sooty-brown plumage, it is well camouflaged in its sylvan surroundings. It nests in lichen-covered branches of mature conifers, and when danger approaches overhead, it lies flat against the nest to blend with the flora. When exchanging incubation duties at night, adults fly within inches of the ground to minimize detection by predators.

Although this bird is abundant from California to Alaska, not a single nest had been recorded until 1959, when one was found on the ground in Alaska. Another was located in 1974, when a down-covered hatchling was found forty meters up a Douglas fir by a California logger. More than a dozen nests have been located since, but because of difficulty in finding them, breeding populations are hard to assess.

In the summer of 1990, researchers in the Walbran Valley, near Vancouver Island's West Coast Trail, found an abandoned marbled murrelet nest. The discovery spurred a controversy in which environmentalists argued against the scheduled clearcutting of the valley, saying the tiny bird, listed as "threatened" in Canada, needs virgin forests for nesting habitat. The marbled murrelet is covered by the Migratory Birds Convention Act of 1916, under which Canada and the U.S. agreed to protect species which cross international boundaries. The Western Canada Wilderness Committee launched a B.C. Supreme Court challenge to the Canadian government to save the forest. The court ruled in favor of the government.

The tufted puffin is considered by many to be the most striking seabird. Its bulky orange beak, shaped like an Indian arrowhead, protrudes from a creamy-white face. It wears a black cap which matches the rest of its body, and blonde tufts extend back from both eyes. Although there are probably thirty-eight thousand nesting pairs in B.C., twelve thousand in Washington, and 1.3 million in Alaska, this clownish-looking bird is seen more in photographs than real life. It nests in burrows on steep, grassy offshore islands, and winters far offshore: it has been seen eight hundred kilometers from land.

Except for its black and white plumage and red feet, the pigeon guillemot looks like an aquatic edition of an ordinary downtown pigeon. It is more likely to be seen near towns and cities than other alcids, as it feeds near shore in harbors and bays, channels and inlets. It inhabits northwest shores throughout the year, and is most abundant from April to September.

These birds are experts at hiding their nests. They lay between one and three eggs in caves, under logs, in burrows, or in rock crevices. As with the marbled murrelet, well-hidden nests make population counts difficult. There are probably twenty-one hundred pairs in Washington, forty-seven hundred in B.C., and twenty-three hundred in Alaska.

Dabblers, Divers and Other Waterfowl

EACH YEAR MORE THAN 6 MILLION migrating ducks, geese, and swans cross international boundaries in the Pacific Northwest. Almost all of them stop to feed and rest on our bays, inlets, harbors, wetlands, and estuaries; thousands stay through winter. Canada and the United States have recognized the significance of the waterfowl we share since the Migratory Birds Convention of 1916.

Through controls on hunting, bird exporting, and nest and egg gathering, the treaty provides protection for our waterfowl, and for millions of seabirds, shorebirds, passerines, and other migratory species. The treaty, however, deals specifically with birds: except for certain pollution regulations, habitats are not covered.

Today, more than seven decades after signing the agreement, 70 percent of the usable habitat in the Fraser delta, one of Canada's most valuable waterfowl regions, has vanished. South of the border, three-quarters of the coastal wetlands in Washington, Oregon, and California are gone. Fertile alluvial floodplains have been ditched and diked to improve their arability; sheltered, accessible shores have become sites for sawmills, log sorting, loading docks, and other heavy industry; flat, open uplands, where roads and services are cheap and easy, have been transformed to residential subdivisions.

Wetland alienation is not restricted to North America. In the early 1970s, wildlife authorities from around the world met at Ramsar, Iran, for the "Convention on Wetlands of International Importance especially as Waterfowl Habitat." Since the Ramsar Convention came into force in 1975, some four dozen countries, including the U.S. and Canada, have pledged to preserve more than four hundred wetlands covering 30 million hectares. Laudable for its intent, the treaty is essentially a gesture of international goodwill, a statement of moral, rather than legal, commitments.

One legal means of wetland preservation in North America is an environmental review. Industrialists and developers are required to submit their proposals to panels of experts to determine if their plans could be detrimental to the environment. If a project would harm or eliminate wetlands or other habitats, the developer could be required to mitigate the damage by preserving another nearby habitat or restoring previously alienated habitats.

BELOW: COMMON GOLDENEYE
WITH BROOD.

The most recent, and possibly the most ambitious, Canada-U.S. agreement, is the North American Waterfowl Management Plan. Designed in 1986, the goal is to involve governments, industries, conservationists, farmers, and individuals in the preservation, enhancement, and maintenance of key waterfowl habitats. Among areas of concern to both countries is the "Middle-Upper Pacific Coast," a stretch of shoreline from San Francisco Bay to the Skeena River, on B.C.'s north coast. Most of this rugged, rocky coastline is of little use to waterfowl, but there are pockets of habitat — estuaries, farmlands, floodplains, freshwater wetlands, rivers, and creeks — that are essential for the production of food, for both wildlife and people.

Proof of their importance is in the numbers of birds that winter on the estuarine mud flats and marshes of the coast. At least 500,000 ducks, 100,000 geese, and more than 15,000 swans winter along the so-called "Middle-Upper Pacific Coast." As many, if not more, scoters and other sea ducks share these areas: some, such as the harlequin duck and oldsquaw, winter nowhere else.

These bird numbers may seem high, but they actually have dropped substantially since the 1970s, as they have elsewhere on the continent. There are exceptions, however: wintering snow geese on the Fraser and Skagit deltas have doubled to forty thousand; Pacific brant at Washington's Padilla Bay are up from three thousand in 1983 to nineteen thousand in 1989. The world population of trumpeter swans, whose future looked grim in the 1930s and '40s, has grown to about ten thousand: a quarter of them winter in the Georgia Strait-Puget Sound area.

The ultimate aim of the waterfowl management plan is to restore migratory bird populations to 1970s numbers by the year 2000. On the B.C. coast, 55,400 hectares of foreshore and uplands would be secured through outright purchases and by giving Crown lands protective designations. On the American northwest coast, most of the 63,600 hectares of wetlands to be secured would be bought from current owners. Total purchase costs for both countries is estimated at $530 million (U.S.). An additional 145,300 hectares could be used by birds if existing rates of habitat losses were stopped, and habitats were enhanced or restored.

Priorities in Canada are the Fraser delta and Vancouver Island's Comox Valley. In the U.S., estuaries and uplands of Washington's Skagit delta, including Debays Slough, and Port Susan and Padilla bays, are considered most important.

Swans and Geese — Vociferous Waterfowl

Swans and geese are the most spectacular, and undoubtedly the most vociferous, waterfowl. To hear them honking overhead, to look up and see them flying in orderly formations above urban neighborhoods, is an agreeable aspect of life, a part of the Pacific Northwest ambience. Their travels up and down the coast are a signal that seasonal changes are upon us. Their daily winter activities draw families to wetlands, seashores, and estuaries to watch, to listen, to learn.

Our largest and, to many, most handsome wildfowl is the trumpeter swan. Weighing over twelve kilograms, its snow-white body measures almost two meters from the tip of its tail to the end of its black beak. There is no bigger water bird on the continent.

ABOVE: TRUMPETER SWANS ARE NORTH AMERICA'S LARGEST WATER BIRDS.

On the coast, these beautiful birds are winter visitors. After producing cygnets in Alaska, they begin to appear in southern B.C., Washington, and Oregon in October. They concentrate in Georgia Strait and Puget Sound, where twenty-five hundred or more spend the winter. About nine hundred winter in the Comox Valley of southeast Vancouver Island. Across the strait, another four hundred or so forage the fields around Washington's Debays Slough, about ten kilometers inland from the mouth of the Skagit River. Several hundred others collect in small groups on sheltered bays, estuaries, flooded fields, and farmlands. They feast in the northwest through winter and begin to move out in late February and early March.

Hunting and habitat losses nearly drove this bird to extinction around the turn of the century. Since the 1960s, however, Alaskan counts of trumpeter swans in late summer have risen from three thousand to about ten thousand.

Trumpeters are occasionally seen with tundra swans, a species once thought to be more abundant on the coast. Perhaps ten thousand or more tundra swans leave Alaskan nesting grounds and migrate through the Pacific Northwest in October and November. A few linger on the B.C. and Washington mainland, but most winter around San Francisco Bay and farther south. They pass through the northwest again from late March to early May. The North American tundra swan was formerly known as the whistling swan.

Trumpeter and tundra swans swim with wings pressed against their bodies, heads held high on extended necks. The smaller mute swan, introduced from Europe, is the bird of ballet fame. It swims with wings arched over its back like elegant sails, its neck gracefully curved and bill bashfully pointed down. It is distinguished from the indigenous swans by a yellow beak with a prominent knob at the base. Despite its daintiness, its size and self-assurance occasionally intimidate small humans: some wild mute swans are fairly tame and don't hesitate to swim up to people strolling on beaches or paddling on lakes.

Even friendlier than the mute swan is the Canada goose. This six-kilogram honker adapts so well to city life that it often crowds out other species — including people — from city parks, golf courses, and favored swimming holes. It nests on top of cedar stumps in downtown Vancouver's Stanley Park, and occasionally sets up house on balconies of highrise apartment buildings. It is common to see large flocks grazing beside airport runways, or in fields next to freeways — places they are not likely to get shot.

This is undoubtedly the most pervasive North American goose, occupying summer and winter ranges across the entire continent. On the northwest coast it nests and winters virtually anywhere there is water and food. In winter, flocks in the hundreds are common along the coast; a thousand have been counted near Masset, on the Queen Charlotte Islands, and as many as five thousand wintering Canada geese have been seen together at Reifel Island, on the lower Fraser River.

A few decades ago this species wasn't as plentiful in the northwest, but efforts to populate vacant habitats have been successful. The Canada goose, like similar species, returns to nest at the place it learned to fly. Biologists have established new populations by transplanting flightless young goslings to new areas. When old enough to breed, they return to these areas to nest.

Coastal migrants move south in October and November, when groups of two thousand and three thousand Canada geese gather on mud flats and fields. Flocks of ten and twelve thousand have been counted in mid-November at Tofino, a major staging area.

The smaller snow goose is the feature attraction in winter in the Fraser delta and Washington's Skagit estuary. A population averaging more than forty thousand in recent years shifts between these two river deltas. They begin to arrive from northern nesting grounds in September and October. The population in the Fraser peaks in November and declines through January, when most of the birds show up in Washington. Northbound migrants arrive in the Fraser-Skagit systems in March.

The geese, which feed on the foreshore and adjacent fields, attract thousands of weekend visitors to the Reifel Bird Sanctuary in the Fraser delta. The striking white plumage of snow geese on both the Fraser and Skagit rivers may be stained a rusty red from mineral deposits in their foraging areas. The snow goose also occurs in a "blue phase," which was once considered a separate species known as the blue goose. Both white and blue snow geese are found in the northwest.

All of these geese breed on Wrangel Island off eastern Siberia, where about 100,000 have nested in recent years. That number is down from about 400,000 in 1960: a goal of the North American Waterfowl Management Plan is to bring the Wrangel Island breeding population up to 120,000 by the year 2000.

Ducks — A Family Affair

Ducks are a youngster's introduction to wildlife. There is hardly a child who hasn't trundled off to the park with a bag of bread for the ducks. Children grow up with ducks quacking and waddling through their songs and storybooks. And when they get older, they take their children and grandchildren — bags of bread in hand — to the local duck pond. Though we call them wildfowl, ducks really are a family affair.

The world is rife with duck lovers. As early as 1887, Canada established North America's first migratory bird sanctuary at Last Mountain Lake in Saskatchewan. Soon after, in 1902, the Audubon Society began working toward the creation of sanctuaries in the United States. Within the next decade, gun and ammunition manufacturers formed the Wildlife Management Institute, which lobbied for the Migratory Bird Convention of 1916.

Hunters are keen wildlife observers, and early in this century it was obvious to them that waterfowl populations were taking a plunge. In 1930 they organized the More Game Birds in America Foundation to promote better management of migratory species, primarily through habitat protection. But money, lots of it, was needed to preserve and enhance waterfowl habitats. In 1937, Ducks Unlimited came into being in the U.S. to raise the necessary funds. Most American ducks, however, breed in Canada, so the following year Ducks Unlimited (Canada) was set up to receive money donated by sportsmen south of the border.

Today, Ducks Unlimited is a privately financed, multimillion-dollar organization with a battery of biologists, engineers, and fund-raisers. Each year DU spends millions to improve waterfowl habitats and increase bird numbers for the enjoyment of sportsmen, birders, children, and grandparents.

About three dozen duck species migrate and winter in the Pacific Northwest. All members of the family Anatidae, they are divided into two main groups. Mallards, pintails, teals, shovelers, and wigeon are dabbling, or puddle, ducks; scoters, goldeneye, buffleheads, and mergansers are diving ducks. These birds are designed to move quickly through water, with streamlined bodies, narrow, pointed wings, and webbed feet. Females are drab compared to the brightly colored males, but all species can be identified by their colored wing patches.

Most ducks nest on the ground and produce a dozen or so eggs. Drakes desert their mates during incubation, joining up with other males of the same

BELOW: A MALLARD HEN, WHOSE BROWN COLOURING RESEMBLES THE MALE'S ONLY WHEN THE DRAKE IS IN ITS "ECLIPSE PLUMAGE."

species and flying off to an isolated swamp or wetland to molt. Ducklings enjoy the protection of their mothers until old enough to fly, but the early life of a duck is harsh, with only 50 percent success on the nesting grounds. Sixty or 70 percent of all ducks fall victim to crows, ravens, coyotes, mink, skunks, racoons, and other predators during their first year. Adults are also vulnerable during the annual molt, a three- or five-week period when they lose all their wing feathers, rendering them flightless.

Puddle Ducks — The Dabblers

When people say "duck" they usually mean "mallard." The most widespread duck in the northern hemisphere, it is the one with the yellow beak, shimmering green head, chestnut breast, and silver sides. It is the one whose "quack" is most familiar.

Two million mallards migrate on the Pacific flyway; forty thousand or more winter in the northwest, where flocks in the thousands are normal. During one Christmas bird count at Ladner Marsh, on B.C.'s lower mainland, more than eighteen thousand mallards were recorded.

Thousands of mallards also remain in the northwest through summer, many to breed. Most nests, made of grass, leaves, sedges, and moss, are built at marshes, lakes, or riverbanks within easy reach of fresh water. About eight or ten olive-green eggs are laid and hatched by the female, who raises the brood without help from the drake.

The mallard's high profile is due not only to abundance, but to adaptability. The agriculture that has taken over much of its natural realm has become an advantage. Mallards contentedly switch from traditional foods, such as sedges,

BELOW: NORTHERN PINTAILS ARE FAVORED BY HUNTERS.

Wigeon and Coots

The most plentiful dabbling ducks on the coast during winter are American wigeon. Perhaps 900,000 migrate down the Pacific flyway and sixty thousand winter on the Fraser delta. Flocks of one thousand and two thousand are scattered along the coast at estuaries, lagoons, marshes, fields, golf courses, and airports.

They are mainly bottom feeders: inland they eat pond weeds, sedges, and grasses; on the sea they feed on eel grass, algae, and molluscs. Wigeon, which dabble in shallow water, keep the company of American coots, which dive deeper. When there's a shortage of food in the shallows, wigeon knock food from the bills of coots and steal it.

Coots, as members of the family Rallidae, aren't ducks, but they frequent the same kinds of places. They drift in rafts of several dozen, on lakes, estuaries, and intertidal marshes — anywhere sufficient aquatic vegetation is available. These are goofy-looking, chicken-sized birds, with heads that bob up and down as they swim, and oversized feet made for plodding across soft mud flats.

seeds, silverweed, insects, and snails, to apples, potatoes, peas, corn, grain, and other domestic crops.

Only in summer, when in its "eclipse plumage," does the drake resemble the mottled brown hen. By autumn the drake is once again dressed in its familiar colors. The mallard is the most important game bird in the northwest, where well over a hundred thousand are taken by hunters each year: some say it is the best-tasting bird of its kind.

Second to the mallard for sportsmen is the northern pintail, a slender, long-necked duck with a sharp, pointed tail. It is almost as plentiful as the mallard on the Pacific flyway, and is often seen on ponds and sloughs with mallards, brant, or wigeon. Migrants from northern nesting grounds arrive first at Puget Sound, in September and October. Soon tens of thousands occupy wintering waters throughout the northwest coast.

About a million pintails use a coastal migration route, traveling as much as 160 kilometers offshore, flying as far down the coast as South America. One banding project showed that pintails wintering in Hawaii had nested in California, Saskatchewan, and Siberia.

Like mallards, thousands of pintails remain near the northwest coast through summer and many breed at sloughs, ponds, lagoons, or lakes. Birds pair up in their wintering areas, then the males follow their mates to the nesting sites, chosen by the hens.

The green-winged teal, one of our smallest waterfowl, likes tidal flats more than

BELOW: COOT WITH YOUNG.

any other duck. It rarely takes to the sea, except to rest on shallow bays and lagoons while migrating. This is a particularly attractive bird, with a green and chestnut head: its lightly spotted, cream-colored breast has a conspicuous white stripe.

On the wing, these birds move in large flocks whose aerial acrobatics, accompanied by the whistling of males, makes identification simple. About 20,000 of the 280,000 that migrate on the Pacific flyway winter in the northwest. They are fair game for hunters, but their size, speed, and erratic flight make individual birds hard to shoot. When they travel in close-knit flocks, however, they are easy targets.

Small numbers of green-winged teals nest on the coast at sloughs and ponds, usually near estuaries. They lay about a dozen eggs in heavily concealed nests lined with down, moss, grass, leaves, and twigs.

Sea Ducks — The Divers

The harlequin duck, with its feathers of blue, rusty brown, black, and white, is unmistakable when swimming among other diving ducks. It is a year-round resident on the northwest coast, where pairs and trios are seen standing on rocks and reefs, preening and loafing. Flocks of a hundred or two hundred

OPPOSITE (UPPER): HARLEQUIN DUCK, THE WOOD DUCK'S SEAGOING RIVAL.

BELOW: THE GREEN-WINGED TEAL IS ONE OF THE NORTHWEST'S SMALLEST WATERFOWL.

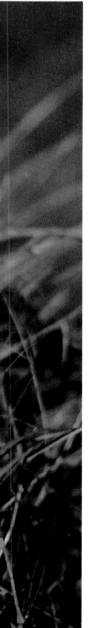

ABOVE: SURF SCOTERS FORM
FLOCKS IN THE HUNDREDS OF
THOUSANDS.

RIGHT: COMMON GOLDENEYE
PAIR.

are common, particularly in turbulent water near rocky shores and islets. It congregates with spawning herring, but feeds mainly around kelp beds on crustaceans, chitons, snails, clams, and small fish.

In early spring harlequins head for nesting areas. A few breed on coastal islands, but most move to rushing streams a short distance inland. Males, which leave their mates to handle domestic duties, return to the sea as early as mid-May, when they select molting sites near rocky shores and islets. Some of the most notable molting locations are in northern Georgia Strait, where groups of a hundred or more males gather from May to August. The normally resplendent drakes are nondescript and flightless during the summer molt. Hens and offspring follow the males to the ocean in August or September.

Like harlequins, surf scoters congregate — in unbelievable numbers — around spawning herring. The most incredible record is from Big Bay, near the mouth of B.C.'s Bute Inlet, where 300,000 surf scoters were counted on April 16 and 17, 1975. Two years later, a flock of 100,000 was surveyed in mid-May at Kinkatla Inlet, on the north coast of B.C. These enormous flocks were likely a mix of wintering scoters and breeders migrating to their northern interior nesting grounds. Up and down the Pacific Northwest coast, flocks in the hundreds and thousands are not unusual.

Often seen with goldeneye, buffleheads, and white-winged scoters, surf scoters are distinguished by their upturned tails and patches of white on their black bodies above the eyes and behind their heads. Their beaks are orange, white and black. These birds dive for molluscs, fish, and crustaceans near beaches and spits, and in lagoons and bays.

BELOW: GREATER SCAUP, OR
"BLUE-BILL."

The larger white-winged scoter, measuring up to sixty centimeters long, is one of the largest ducks in the northwest. It prefers deeper water than the surf scoter and is seen up to five kilometers offshore. With brown-black plumage, white wing and eye patches, and orange-yellow bill, it is not as colorful as the surf scoter, but no less interesting to watch. Like the surf scoter, the white-winged

LEFT: BUFFLEHEAD, ONE OF
NORTH AMERICA'S SMALLEST
DUCKS.

scoter nests in the interior. Flocks of nonbreeders in the hundreds and thousands, however, inhabit the coast through summer.

The common goldeneye is another herring lover that mingles with other divers during spring spawn. Though not as abundant as the scoters, the goldeneye fishes similar areas, and rafts of a few hundred are normal. Its colloquial

Redhead, the Coastal Birder's Challenge

An uncommon, but pleasant sight on the northwest coast is the redhead, mainly a winter visitor to the Georgia Strait-Puget Sound area, where it appears alone or in groups of a few dozen.

Small numbers are scattered along the coast in summer, but the largest year-round concentrations occur inland in southern B.C., and northern Washington and Idaho. On the coast it feeds on vegetation at lakes, saltwater lagoons, and flooded fields.

The redhead is a particularly attractive bird, with a large, chestnut head, black breast and tail, and silver-gray body. Its pale blue beak has a shiny black tip. The redhead's infrequency at the seashore presents a challenge to keen coastal birders.

name, "whistler," is a reference to the high-pitched whistling of its wings as it flies over the surface of the sea. A few remain on the coast through summer, but breeders move to northern and interior woodlands to build nests in woodpecker holes and other cavities.

Another cavity nester that winters on the coast is the bufflehead, one of North America's smallest ducks. It is readily identified by its well-feathered head, bearing a large white patch behind the eyes. This bird is most often seen in small groups of a dozen or so, but flocks in the hundreds, even thousands, are occasionally sighted, particularly around spawning herring. Most leave the coast by April or May, and the species is rarely seen on the ocean again until October or November.

The greater scaup, or "blue-bill" is somewhat similar to a common goldeneye, with bright yellow eyes and metallic green-black head, but its beak is a pale grayish-blue. Like other sea ducks, this bird enjoys herring, but also feeds on shellfish, crabs, insects, and plants. On the coast it is just as likely to be seen on lagoons, harbors, and inlets as on lakes, rivers, or flooded fields.

A few greater scaups spend summer on the coast. In March, most begin moving toward breeding areas in the far north, where they build ground nests near fresh water. In October and November, they return to northwest seas, where probably twenty thousand or more stay for winter.

Loons and Grebes

People camped on the shores of the Pacific Northwest may be haunted through the night by the ghostly laughter of the loon. Some believe it's the call of a maniac — crazy as a loon, they say — a persistent, melancholy cry that pierces the still twilight before sunrise.

The loon never seems to sleep: its restless cry, as singular as the howl of a wolf, carries on through the night. For some people it is an unsettling sound: for others it's a soothing companion in the wilderness. Regardless of how the cry of the loon arouses the imagination, few would deny it is the true call of the wild.

Four loon species — the red-throated, yellow-billed, common, and Pacific — inhabit the northwest coast at various times. The red-throated loon is often seen in winter and during migrations, gathering in flocks of fifty or fewer. The yellow-billed loon, distributed widely along the coast, occurs in small numbers throughout the year.

The loon frequently depicted in Indian art and legend is the common loon, clad in striking summer plumage, with a velvet black head and string of pearls around its neck. It looks as though someone tossed seashells, bleached by the sun, onto the bird's back: according to native folklore, that's precisely what happened.

The common loon is a big diving bird, sometimes nearly a meter long, and it lies low in the water, cruising around in search of fish. Its name may be misleading in the northwest, because it's not the most common loon here. Its smaller cousin, the Pacific loon, occurs in huge flocks migrating in spring through Juan de Fuca and Georgia straits to its northern nesting grounds.

The Pacific loon is the most gregarious, gathering by the thousands in areas where tidal upwellings stir up feasts of herring, sandlance, or anchovies. One of these places is Active Pass, in the Gulf Islands: in April, passengers aboard B.C. Ferries may

Fish Ducks

Mergansers have long, serrated bills, designed to grab fish on the move. Known as "fish ducks," the three merganser species on the northwest coast are handsome birds that are widely distributed, but not abundant. The red-breasted merganser is more seagoing than the common or hooded merganser, and is likely to be seen in deep water. It is somewhat of a socialite among ducks that swim tidal waters. The common merganser, also a sociable bird, prefers lakes, rivers, or brackish and near-shore water, with rocks, logs, sand bars, and shoals for resting and roosting.

The hooded merganser, smallest of the three, frequents estuaries, lakes, sloughs, and streams. It is one of the least gregarious diving ducks, rarely seen with more than about fifteen birds of the same species.

see as many as ten thousand Pacific loons fishing the turbulent waters of the pass.

When fishing, the loon dives to great depths, propelled by powerful paddle-like feet. There are records of loons being caught in fish nets sixty meters below the surface. They dive forward in a quick roll, exposing the tips of their stubby wings for a split second before disappearing. Sometimes a loon will expel the air from its lungs and simply sink from sight, leaving no trace of having been there. Loons have large amounts of myoglobin in their muscles; this respiratory pigment allows them to store oxygen for use underwater.

By early or mid-May, most Pacific loons are migrating toward their breeding grounds in northern B.C. and beyond. The enormous flocks which move up the coast are impressive: during the latter half of May, more than 150,000 loons are known to fly past Brooks Peninsula, on the northwest coast of Vancouver Island. Although the loon has difficulty taking off, running upwind along

the surface of the water, it is a powerful flier once it's airborne, covering vast distances at speeds up to a hundred kilometers an hour.

As they reach the north, the flocks split up and the birds settle on forested lakes or ponds. They are solitary nesters, returning each year to breed in the same areas. The loon rarely leaves the water except to fly and breed, and it is during the nesting season that the unfortunate loon's ponderousness becomes difficult to conceal. The loon was designed to swim, with legs set so far back that when it hauls itself from the water it often flops forward on its breast when it tries to move too quickly. The word loon comes from a Scandinavian word, *lom*, meaning lame or clumsy person. It is rather unfair when you consider the loon's enviable efficiency in water.

The northern nesting grounds are usually in isolated, uninhabited places, and it is here that the loon has earned its reputation as a symbol of true wilderness. Thousands of loons winter in the northwest, some as far south as

California, but the loon is inevitably associated with the loneliness and tranquility of the far north.

The loon's coastal fishing waters are shared with the western grebe, another large diving bird that nests inland. Like the loon, it gathers by the thousands where feed is bounteous, especially in late winter and spring when herring are spawning. Flocks of eleven thousand have been counted in March off southern Vancouver Island.

Unfortunately for coastal wildlife watchers, the grebe exhibits its most fascinating behavior on its freshwater breeding grounds. When in amorous pursuit of a mate, this bird walks on water: in an elaborate maneuver called "rushing," grebes splash across the surface side by side, then dive simultaneously. This courtship dance may be carried out by several pairs at a time.

In nests on the water, hidden among tules or rushes, both sexes care for a brood of three or four. Freshly hatched "grebelings" immediately climb into a pocket formed by the back feathers and wings of the adult. The nest is soon abandoned and the birds move onto the marsh. While one parent carries the young on its back, the other gathers the family meals.

The western grebe's scientific name, *Aechmophorus occidentalis*, meaning "western spear-bearer," is descriptive of its long, rapierlike bill. Though there is no scientific proof, it is widely believed that grebes spear fish.

OPPOSITE: THE COMMON MERGANSER, ONE OF THREE MERGANSER SPECIES ON THE NORTHWEST COAST.

BELOW: HOODED MERGANSER, SMALLEST OF THE "FISH DUCKS."

METRIC CONVERSIONS

This book conforms to the metric system, which is often confusing to people unaccustomed to dealing with it. But simple conversions, such as inches to centimeters, miles to kilometers, or pounds to kilograms, present little problem when you know the multiplication factors. The following table outlines the conversion factors.

When you know	Multiply by	To find
centimeters	.4	inches
meters	3.3	feet
kilometers	.63	miles
square meters	1.25	square yards
square kilometers	.4	square miles
hectares	2.5	acres
kilograms	2.2	pounds

Or:

inches	2.5	centimeters
feet	.3	meters
miles	1.6	kilometers
square yards	.8	square meters
square miles	2.6	square kilometers
acres	.4	hectares
pounds	.45	kilograms

BIBLIOGRAPHY

Baird, R.W. "Elephant Seals on the Coast." *Animals*. B.C. SPCA, 1990.

Beebe, F.L. "Field Studies of the Falconiformes of British Columbia." Victoria: B.C. Provincial Museum, 1974.

Bigg, Michael A. et al. *Killer Whales*. Nanaimo, B.C.: Phantom Press & Publishers Inc., 1987. "Status of Steller Sea Lions and California Sea Lions in British Columbia," 1985; "Social Organization and Genealogy of Resident Killer Whales in the Coastal Waters of British Columbia and Washington State," 1988; "Migration of Northern Fur Seals off Western North America," 1990. Ottawa, Ontario: Department of Fisheries and Oceans.

Breen, Kit Howard. *The Canada Goose*. Vancouver, B.C.: Whitecap Books Ltd., 1990.

Butler, Robert W. and Campbell, Wayne. "The birds of the Fraser River Delta: populations, ecology and international significance." Canadian Wildlife Service, 1987.

Campbell, R. Wayne et al. *The Birds of British Columbia*. Victoria, B.C.: Royal B.C. Museum, 1990.

Cottingham, Dave and Langshaw, Rick. *Grizzly*. Banff, Alta.: Summerthought Publications, 1981.

Darling, Jim. "Gray Whales off Vancouver Island," 1984; "Survey for Nestucca Oil Sediments of Selected Gray Whale Feeding Sites," 1989. Ottawa, Ont.: Department of Fisheries and Oceans.

Davidson, Art. *In the Wake of the Exxon Valdez*. San Francisco, CA: Sierra Club Books, 1990.

Dorst, Adrian and Young, Cameron. *Clayoquot: On the Wild Side*. Vancouver, B.C.: Western Canada Wilderness Committee, 1990.

Fisheries and Oceans, Dept. of. "Report of the Fisheries and Oceans Research Advisory Council on Seal Research in Canada," 1987; "Guidelines for Whale Watching," 1986. Ottawa, Ont.: Department of Fisheries and Oceans.

Fitzharris, Tim. *The Island: A Natural History of Vancouver Island*. Toronto, Ont.: Oxford University Press, 1983.

Francis, Daniel. *A History of World Whaling*. Markham, Ont.: Penguin Books Canada Ltd., 1990.

Guiguet, C.J. *The Birds of British Columbia (series): The Crows and Their Allies*, 1954; *Waterfowl*, 1967; *The Shorebirds*, 1955; *Gulls, Terns, Jaeger and Skua*, 1957; *Diving Birds and Tube-nosed Swimmers*, 1982. Victoria, B.C.: B.C. Provincial Museum.

Haley, Delphine et al. *Marine Mammals of Eastern North Pacific and Arctic Waters*, 1986; *Seabirds of Eastern North Pacific and Arctic Waters*, 1984. Seattle, WA: Pacific Search Press.

Hamilton, A.N. and Nagy, J.A. "Working Plan, Khutzeymateen Valley Grizzly Bear Study." Victoria, B.C.: Ministries of Environment and Forestry, 1990.

Herger, Bob and Neering, Rosemary. *The Coast of British Columbia*. Vancouver, B.C.: Whitecap Books Ltd., 1989.

Hoyt, Erich. *Orca: The Whale Called Killer*. Camden East, Ont.: Camden House Publishing, 1990.

Islands Protection Society. *Islands at the Edge: Preserving the Queen Charlotte Islands Wilderness*. Vancouver, B.C.: Douglas & McIntyre Ltd., 1984.

Kenyon, Karl W. *The Sea Otter in the Eastern Pacific Ocean*. Washington, D.C.: U.S. Government Printing Office, 1969.

BIBLIOGRAPHY

Leatherwood, Stephen et al. *Whales, Dolphins, and Porpoises of the Eastern North Pacific and Adjacent Waters*. New York, NY: Dover Publications, Inc., 1988.

MacAskie, Ian. *The Long Beaches, A Voyage in Search of the North Pacific Fur Seal*. Victoria, B.C.: Sono Nis Press, 1979.

McConnaughey, Bayard H. and Evelyn. *The Audubon Society Nature Guides, Pacific Coast*. New York, NY: Alfred A. Knopf, Inc., 1985.

McTaggart-Cowan, Ian and Guiget, C.J. *The Mammals of British Columbia*. Victoria, B.C.: B.C. Provincial Museum, 1956.

Nickerson, Roy. *Sea Otters, A Natural History and Guide*. San Francisco, CA: Chronicle Books, 1984.

Obee, Bruce. *The Pacific Rim Explorer*, 1986; *The Gulf Islands Explorer*, revised 1990. Vancouver, B.C.: Whitecap Books Ltd. "Seal-Salmon Controversy Escalates in B.C." *Oceans*, 1987. "Cougar in B.C." 1983; "Wolves of B.C. Predator or Prey?" 1984. *Wildlife Review*. "Sea Otters Return," 1986; "Snow Bear That Triggers the Imagination," 1989. *Beautiful British Columbia*. "Oil Spill Aftermath," 1989; "Strip Mining the Seas," 1990; "Gentle Giants," 1991. *Canadian Geographic Society*.

Olesiuk, Peter, et al. "Life History and Population Dynamics of Resident Killer Whales in the Coastal Waters of British Columbia and Washington State," 1988; "Seals and Sea Lions of the British Columbia Coast," 1990; "Recent Trends in the Abundance of Harbour Seals in British Columbia," 1990. Ottawa, Ontario: Department of Fisheries and Oceans.

Peterson, Roger Tory. *A Field Guide to Western Birds*. Boston MA: Houghton Mifflin Company, 1969.

Reader's Digest. *Outdoors Canada*. Montreal, P.Q.: The Reader's Digest Association Ltd., 1977.

Robertson, Kathleen. "Racoons." *Wildlife Review*, 1984.

Scammon, Charles M. *The Marine Mammals of the Northwestern Coast of North America*. San Francisco, CA: John H. Carmany and Company, 1874.

Thomson, Richard E. *Oceanography of the British Columbia Coast*. Ottawa, Ont.: Department of Fisheries and Oceans, 1981.

Trites, Andrew W. "Northern Fur Seals: Why Are They Declining?" Vancouver, B.C.: Department of Zoology, University of British Columbia, 1990.

Udvardy, Miklos D.F. *The Audubon Society Field Guide to North American Birds*. New York, NY: Alfred A. Knopf, Inc., 1977.

Vermeer, Kees et al. "The ecology and status of marine and shoreline birds in the Strait of Georgia, British Columbia," 1987; "Pelagic Seabird Populations off Southwestern Vancouver Island," 1987; "Pelagic Seabird Populations in Hecate Strait and Queen Charlotte Sound: Comparison with the West Coast of the Queen Charlotte Islands," 1984; "Status of Nesting Seabirds in British Columbia," 1984; "Influence of Habitat Destruction and Disturbance on Nesting Seabirds," 1984. Canadian Wildlife Service.

INDEX

ABOUT THE AUTHOR
AND PHOTOGRAPHER

Bruce Obee, a veteran Vancouver Island writer, has specialized in outdoors, travel, wildlife, and environmental topics since 1977. Winner of several journalism awards, his articles are published by *Canadian Geographic*, *Beautiful British Columbia*, *Nature Canada*, *Travel & Leisure*, the *New York Times*, *Alfred Hitchcock's Mystery Magazine* and others. His writing also appears in high-school and university textbooks. He lives in Deep Cove, on Vancouver Island's Saanich Peninsula, with his wife, Janet Barwell-Clarke, and daughters, Lauren and Nicole.

Tim Fitzharris is author and photographer of nine books on natural history, including *Canada: A Natural History*, *The Audubon Society Guide to Nature Photography*, and most recently, *A Journey into North America's Vanishing Forests*. His articles and pictures have appeared in many leading nature magazines, including *National Geographic*, *Audubon*, *National Wildlife*, *Orion*, *Equinox*, and *The Living Bird Quarterly*. His photographs have been exclusively featured in more than a dozen calendars and have been exhibited by the National Museum of Canada and other galleries. He lives in Bellingham, Washington.

BY THE SAME AUTHOR AND

PHOTOGRAPHER

By Bruce Obee

The Gulf Islands Explorer

The Pacific Rim Explorer

By Tim Fitzharris

The Island: A Natural History of Vancouver Island

The Wild Prairies: A Natural History of the Western Plains

The Adventure of Nature Photography

British Columbia Wild: A Natural History

Wildflowers of Canada (with Audrey Fraggalosch)

Canada: A Natural History (with John Livingston)

Wild Birds of Canada

The Audubon Society Guide to Natural Photography

Forests: A Journey into North America's Vanishing Wilderness